Michael Mandiberg

Timeframe

Table of Contents

labor/record
by danilo machado 4

Live Study
Plates 20

Live Study — Performing the Artist Studio
by Christiane Paul 48

The Zoom Paintings
by Michael Mandiberg 60

Zoom Paintings
Plates 70

Curriculum Vitae 97

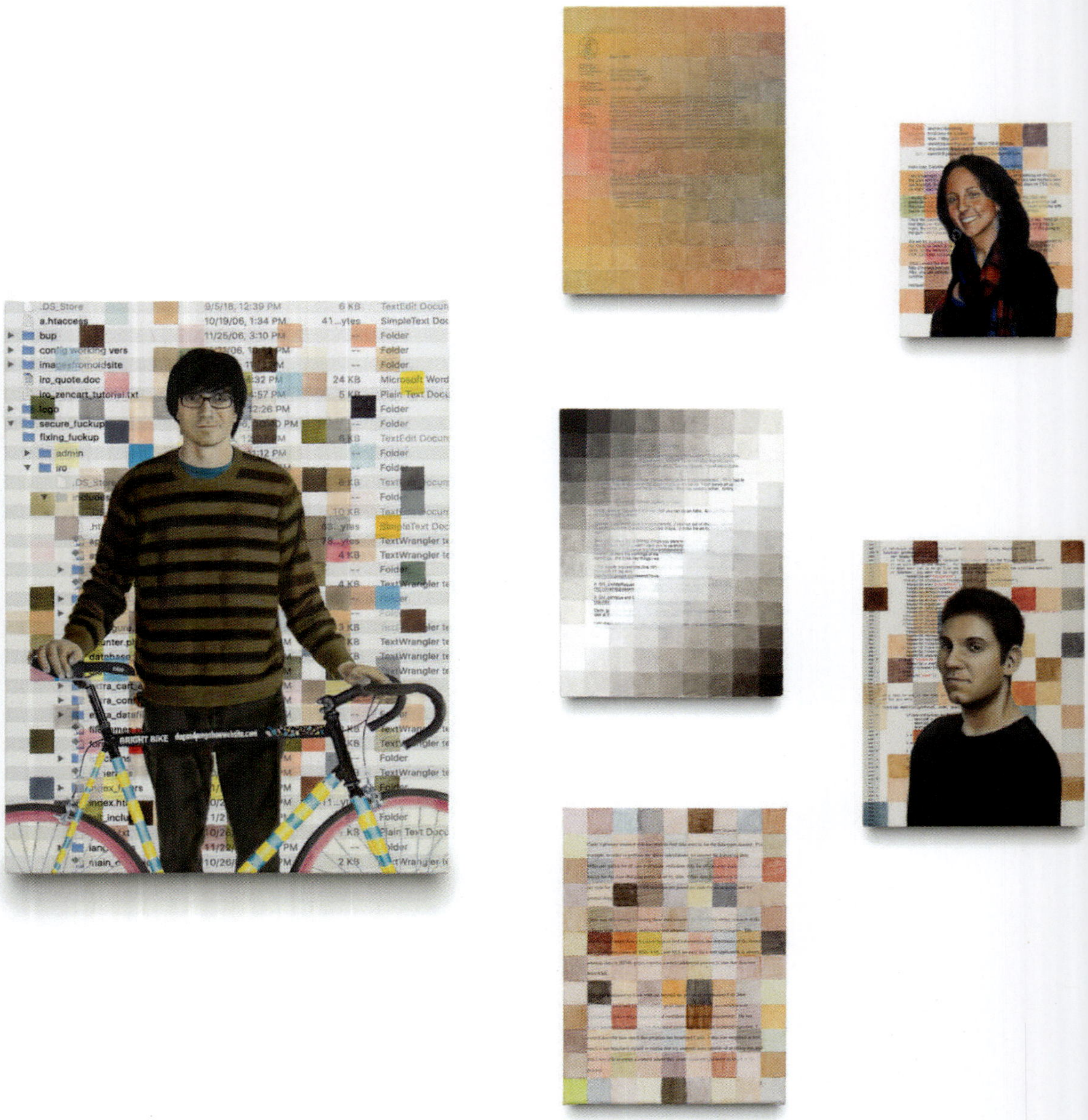
.DS_Store
a.htaccess
bup
iro_quote.doc
logo
secure_fuckup
fixing_fuckup
admin
iro
Folder
TextWrangler
BRIGHT BIKE

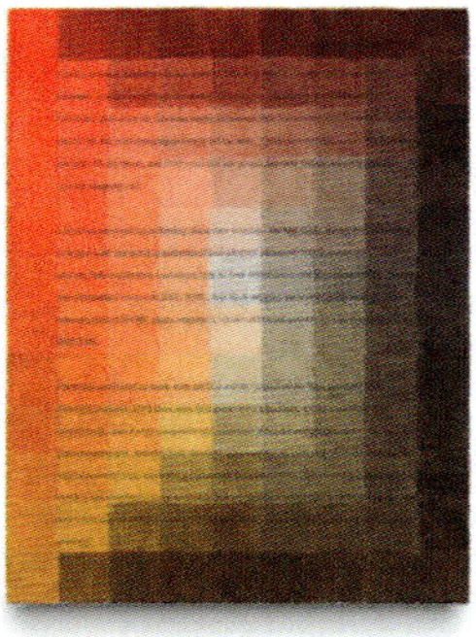

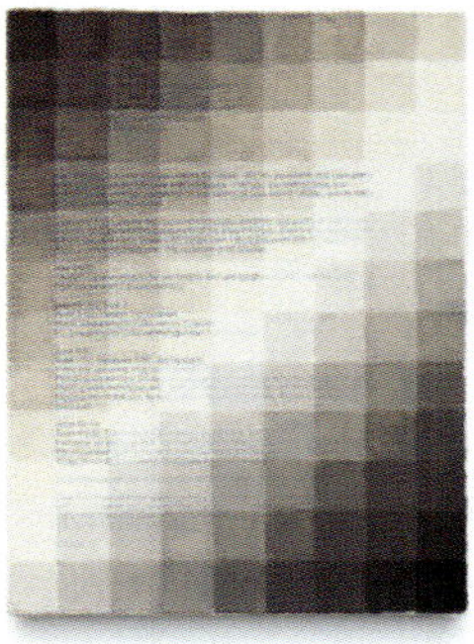

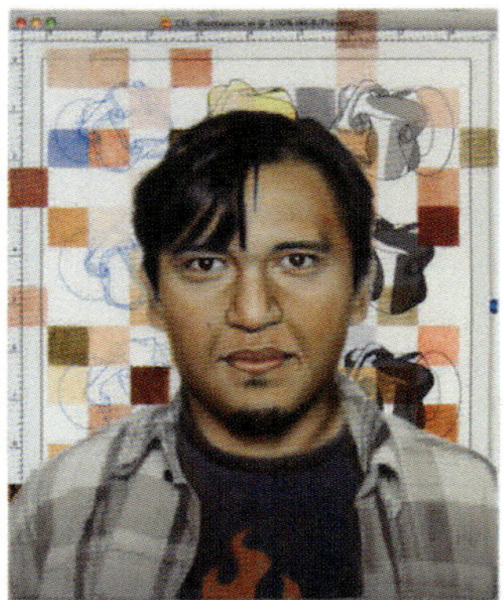

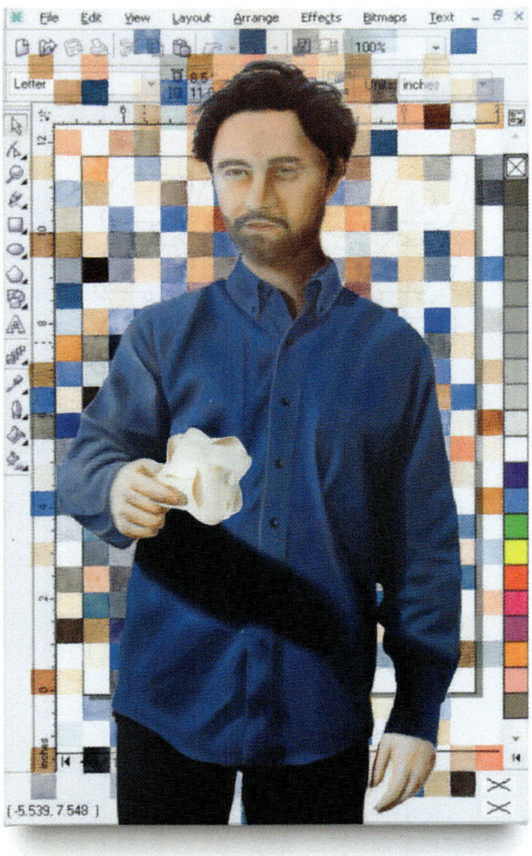
File Edit View Layout Arrange Effects Bitmaps Text
100%
Letter
inches
(-5.539, 7.548)

labor/record

danilo machado

Through careful duration studies, Michael Mandiberg's practice asks how one makes labor visible. Their work refuses binaries, locating itself inexactly between the bodily and the virtual, between portrait and performance. Mandiberg continues their recordings of time and relationships in two recent series, *Live Study* and the *Zoom Paintings*.

Like many during the pandemic, I have been staring at a seemingly infinite number of video rectangles with toggling focus for months and months, making the settings depicted in Mandiberg's *Zoom Paintings* distinctly familiar. The shelves, windows, and furniture in the oil paintings make me think about my own shifting backgrounds—an assortment of art, books, and plants that have been visible behind me during calls this past isolating year. Within the ongoing experience of gaining intimacy with these digital rooms, viewing the set of paintings in person felt uncanny. In part because of my own household's immunocompromised status—one shared by the artist—seeing anyone outside of the Zoom squares had only recently felt possible for me at the time of my visit to Michael's midtown studio. Now, as I am writing this, everything feels increasingly like a risk again as our pandemic reality persists.

Facing page:
Detail

Michael + Taylor re: Artnet, 5:30 — 6:30 PM, November 4, 2020 (#32), 2020

Oil on canvas
6h x 11w in

While our two bodies were present in the studio that July afternoon, there are no bodies in these paintings. For Mandiberg, their omission emphasizes an "interchangeability" between the scenes, which vary in intimacy and anonymity (though the two aren't always opposites). In some, the viewer sees the depths of living rooms and hallways, in others just a cloudy wall color or ceiling vent. While during some larger Zoom calls the

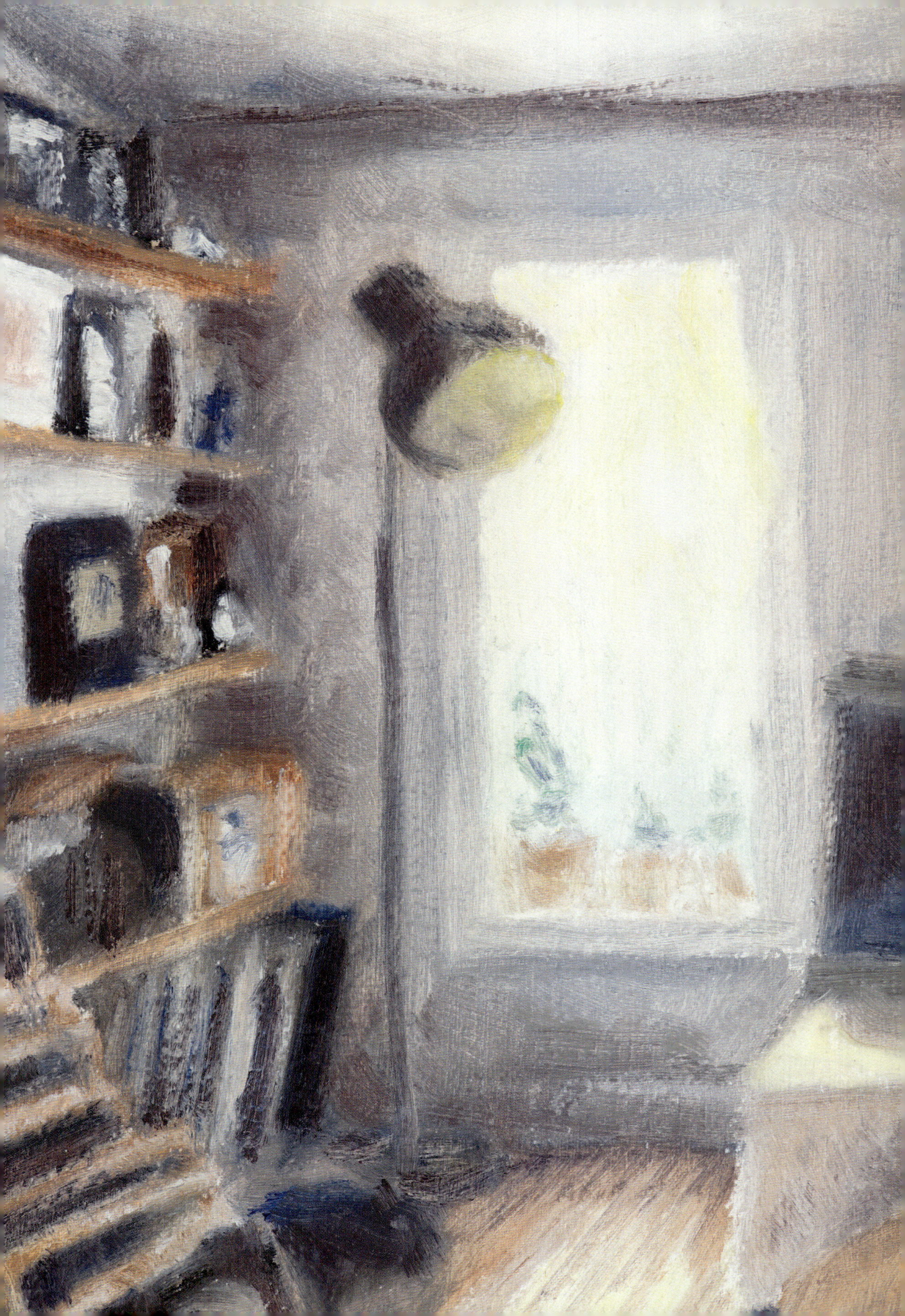

artist chooses which frame to paint, the series' content is unarranged and uncomposed. In fact, many of the canvases break traditional rules of composition and balance. It's a sketchy, queer kind of painting not so much interested in technical mimcry or legibility, but in an archive of time Mandiberg shares with other people on screen, or, in some cases, time reclaimed by the artist.

Describing their process as "memorializing how unmemorable it is," the *Zoom Paintings* recall other

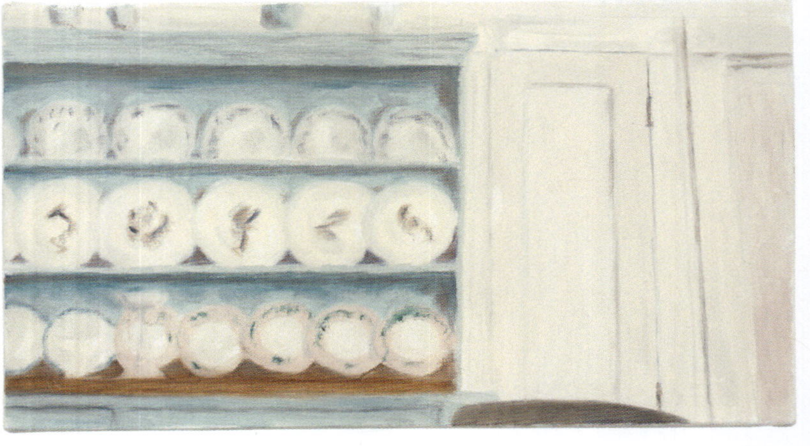

***French Landscape at the Margins of Survival, 1:00 — 2:00 PM, August 10, 2020 (#22),* 2020**

Oil on canvas
6h x 11w in

projects like Gordon Hall's photo book *Other People's Houses* (published by A.I.R in collaboration with Walls Divide Press and fundraising for Project EATS), as well as two earlier works by Mandiberg. In *View from the Window at Work* (2016-17), Mandiberg paid hundreds of Amazon Mechanical Turk workers to take photos out of the window of the room they worked at and displayed the 4 by 6 inch images in a large grid. The *Zoom Paintings's* compilation of different kinds of working windows also feels like a kind of continuation of the (un)memorable

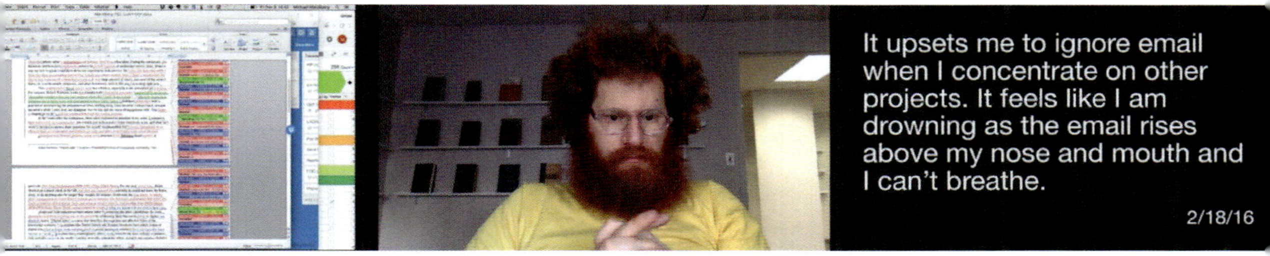

images captured every fifteen minutes by technology used to track freelance labor in *Quantified Self Portrait (One Year Performance)* (2016-17)—but also like a softer, less precise inverse. The *Zoom Paintings*, echoing the compositional ratio and grid display of the platform, capture some of the cramping of quarantine. Two paintings from the series—*French Landscape at the Margins of Survival, 1:00 — 2:00 PM, August 10, 2020* and *Emergency CUNY Budget Rally Online TOMORROW!, 9:00 — 10:00 PM, April 1, 2021*—stand out both because of their class depictions and their conjurings

Above:

Still from ***Quantified Self Portrait (One Year Performance)***, 2016-17, 3 Channel Video

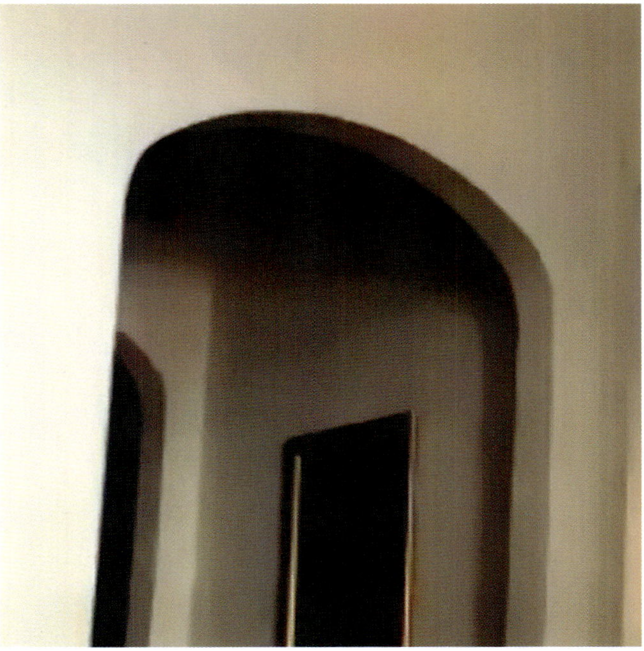

Image from: Gordon Hall, ***Other People's Houses.***

Zine published by Walls Divide Press / A.I.R. Inkjet printed/Staple bound,
21 pages. 2021

of the many. The china cabinet of delicate plates and shelf of red picket signs rendered in the works become reminders of how we have renegotiated gathering during this period.

The *Zoom Paintings* continue many of the durational themes of *Live Study* (2018-present), a performance in which the artist creates portraits of assistants, apprentices, and students over a span approaching 1,000 recorded (and sometimes livestreamed) hours, and counting. Elements reflective of their work overlay the subjects of these paintings: lines of code, the frames of program windows, folder names, emails. In some, tiny profile pictures are visible in the headings of messages and document comments. Mandiberg makes two of each painting: one to gift to the portrait's subject and one to sell. For Mandiberg, it is important that only their hands produce the works, electing oil paint over photography to echo the learning process that takes place with their students in the studio. (Another medium would be "too easy," the artist tells me.) Mandiberg tests color, value, and line in 9 by 12 inch studies composed of many squares and layers of transparent glazes, then incorporates these gridded motifs into the background of some of the portraits. Only having taken one undergraduate painting class, the artist tells me that "getting it wrong is part of getting it right, and getting better is a part of the work." The squares are reminiscent of another ongoing portrait project, *Synecdoche* by Byron Kim. Since 1991, Kim has painted the skin color of friends, family, and fellow artists on wooden panels, groupings of which are installed in infinite configurations. Beyond the technical echoes, Mandiberg's *Live Study* paintings share a sense of accumulation and also consider the ways parts stand-in for the whole as conjured by Kim's title.

Facing page:
Detail

***Oscar Torres (180 hours, 2008)*, 2020**

Oil and inkjet on canvas
15h x 12w in

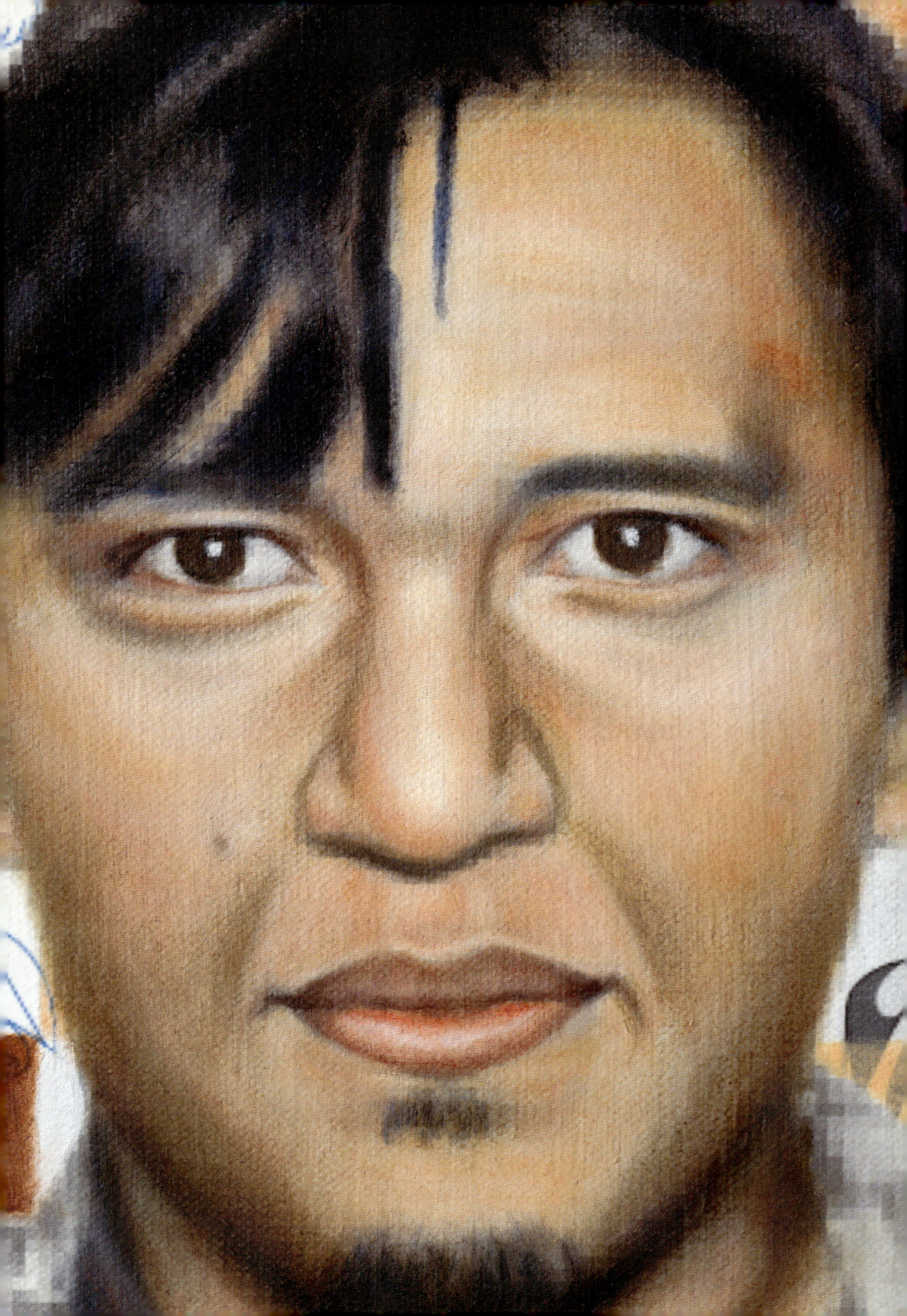

Both *Live Study* and the *Zoom Paintings* narrate an economy of time. The *Live Study* paintings are priced at the total wages paid by the artist to the subject. Their dimensions,

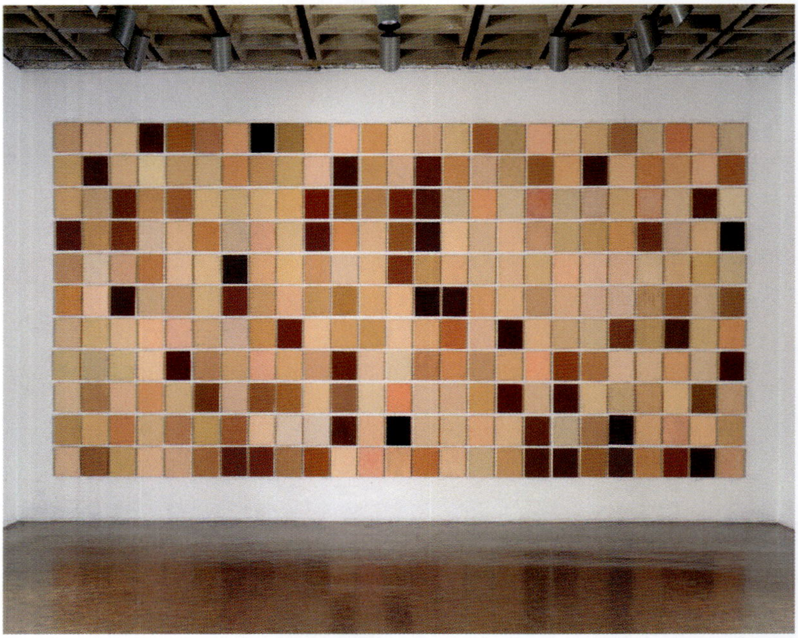

Byron Kim
Synecdoche
(1991-present)
oil and wax on lauan plywood, birch plywood, and plywood

Image courtesy of the artist and *James Cohan*, New York.

which includes canvases measuring 4 by 5 inches and 4 by 5 feet, correspond with the hours the subject spends in the studio working, one square inch per hour. (Time is slippery, and these are not the total hours the subject has spent with the artist, since with some, the artist shares many more hours, whether in class settings or in non-work contexts.) Several paintings show time sheets, requests for stipends, and letters of reference with varying degrees of legibility. They recall the work of artists like Jenny Holzer and Sadie Barnet, who, in distinct contexts, reproduce "official" documents in their paintings and question the role of evidence and authority. For Mandiberg, these documents do not just reflect the bureaucratic or procedural, but also

Facing page:
Detail

Patrick Davison (737 hours, 2008-2010), 2021

Oil and UV print on canvas
31h x 24w in

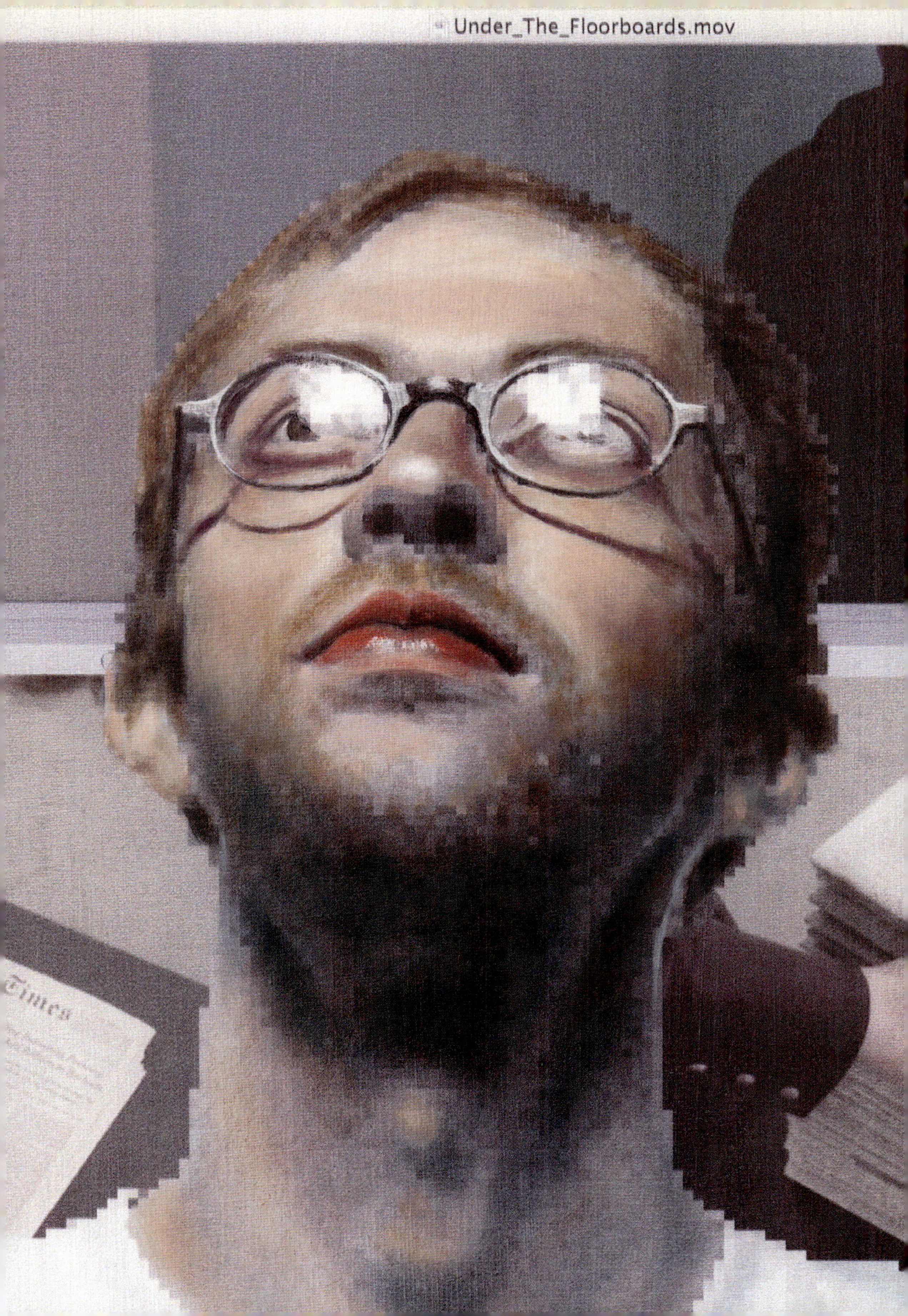
Under_The_Floorboards.mov
Times

the carework and tutelage enveloped within it. Indeed, the titles of two test canvases praises how "Carlo has grown immensely." The artist's choice to represent subjects they have lost touch with in single blurred black and white paintings, such as *Jonathan Perlman (180 hours, 2010)* (2020), further emphasizes a gradient of relationships represented in the series.

While the *Zoom Paintings* are equal in size, they each denote different kinds of time. Their titles are composed of the calendar entry, hour, and duration of each call, alluding to a range of Zoom experiences from the personal (*Uncle Bob 85th birthday via Zoom, 3:00 — 4:00 PM, August 16, 2020*) to the professional (*History of Art All-Staff Introductions, 4:30 — 5:00 PM, September 23, 2020*); from one-on-one conversations with family to participation in webinars among many other attendees. What becomes text in both series (but particularly in *Live Study*) are the ways artists, students, faculty, and educational institutions relate to and rely on each other, and more specifically on each other's labor. This reliance is reflected in both individual relationships and at the institutional level, where organizing around working conditions and pay remains urgent—particularly in the face of compounding pressures of the pandemic. In (and between) both levels, this interdependency is often shaped by inequitable power structures, yet can also be framed by care. As Mandiberg practices care through apprenticeship, their students reciprocate it. The students' support often came in the form of body doubling for the artist during periods of illness when Mandiberg was physically unable to complete their work by themselves. As Mandiberg practices care through apprenticeship, their students reciprocate it—sometimes through extensive physical support/ hands on

Facing page:
Detail

The surgery was more intense than I or the doctors expected (Transparent White vs Titanium White), 2021

Oil and inkjet on canvas
12h x 9w in

<dmpalladino@gmail.com>, Michele Rose <dchallen@gmail.c
Oscar Torres <ot289@nyu.edu>, Tobi Leingruber <tobi@tobi-

Hey, so the surgery was more intense than I or the doctors expected. T
go pretty deep to get one of the lymph nodes in my pelvis. I can barely s
So needless to say, I wont be in on Monday. Prob not tuesday either. Ai
weds.)

DMP, you will be in on monday to work on /labs?

Oscar, work w/ Danielle if she has stuff you can do on /labs. And get the
done for the CFL shade.

Michele, keep working on the screenshots. if you run out of chapters, le
i'll point you in the direction of the web chaps. (i know the ex fo
ing.)

Tobi, you have a set of GM/xpi things you were w
through them (or hit a wall) i want you to carefully
http://transition.turbulence.org/Works/oilstandard/
function to trace the workings of the
standings. the three key things i wa

1 the regular expressions (line 161-
more info on reg ex's:
http://www.google.com/search?q=ia

2. GM_xmlhttpReques
http://diveintogreasem

3. GM_getValue and G
http://div

Carlo, g
look at th

i will chec

labor, particularly essential during periods of illness when the artist was physically unable to complete their work by themselves. The *Live Study* paintings become portraits of how we all—but specifically how Mandiberg as a disabled person and educator—rely on one another to navigate and create within exclusionary systems.

The *Zoom Paintings*, ten of which were exhibited virtually on the same video platform by Denny Dimin in November 2020, make me consider the moments where the corners of

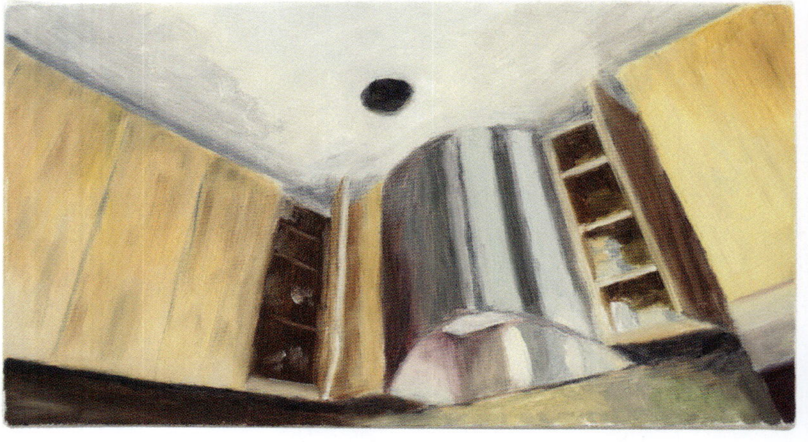

Uncle Bob 85th birthday via Zoom, 3:00 — 4:00 PM, August 16, 2020 (#24), 2020

Oil on canvas
6h x 11w in

the Zoom room seem to disappear, where the connection on a call felt electric, haptic. I think about the use of virtual backgrounds and blurring effects making the squares both feel more in a place and more outside of one. I also think about the disability-centered parties of the pandemic, particularly the Remote Access series, curated by the Critical Design Lab. Described as "crip nightlife praxis" in the detailed and evolving participation guides, the gatherings feature a range of musical and movement performances and incorporate rotating access roles, including captioners and sound describers. I remember the first party in the Spring of 2020, when all of us in the apartment (puppy included) put on cheap wigs and danced among pages of participants. I believe we were logged in twice, one of us on our phone moving around and the other on my desktop which, at the time, lived on our kitchen table. It felt like another glitch, a rupture where not only the window of the video screens were broken but our own walls collapsed outward and engulfed the spaciousness of the many many other rooms. This expansiveness honored

History of Art All-Staff Introductions, 4:30 — 5:00 PM, September 23, 2020 (#29), 2020

Oil on canvas
6h x 11w in

and continued the ways sick and disabled folks like ourselves have organized and gathered online, imagining infinitely more access points than the capitalist platform was originally built for.

While disabled folks were often previously denied access to even basic virtual accommodations, some crips were now the most equipped for the abrupt digital switch that occurred for many last year. There will be no "post" pandemic, but I am curious how this (sometimes fickle) relationship to the virtual will continue morphing, and if those communities will be as ignored again as quickly as they were suddenly tapped for their insights. How will access to digital space for work and pleasure be valued? How will access to this space continue to shift the ways we consider authorship, collaboration, and flexibility? What parts of the infrastructures we have built to engage, to share, and—like in *Quantified Self Portrait*—to be tracked digitally remain commonplace?

Mandiberg's work overlaps painting, portraiture, and performance beyond any clean binaries. It ponders the relationships between care and work, encouraging the viewer to consider how we relate to one another while surrounded by institutions and pandemic. What kind of time do we spend together and how do we remember or record it? How is the slippery experience of pandemic time related to sick time? In its expansive reflections of time, labor, and archive, the artist's practice provides entry points to these questions and more.

Facing page:
Detail

Simon Jolly (636 hours, 2007-2009), 2019

Oil and inkjet on canvas
32h x 20w in

Deep gratitude to Michael for your care-filled work, to Elizabeth, Julia, Kuldeep and Denny Dimin for tending the process of this essay so tenderly, and to my many support structures, especially Malcolm, Ezra, Ruggy, and Claire.

- danilo

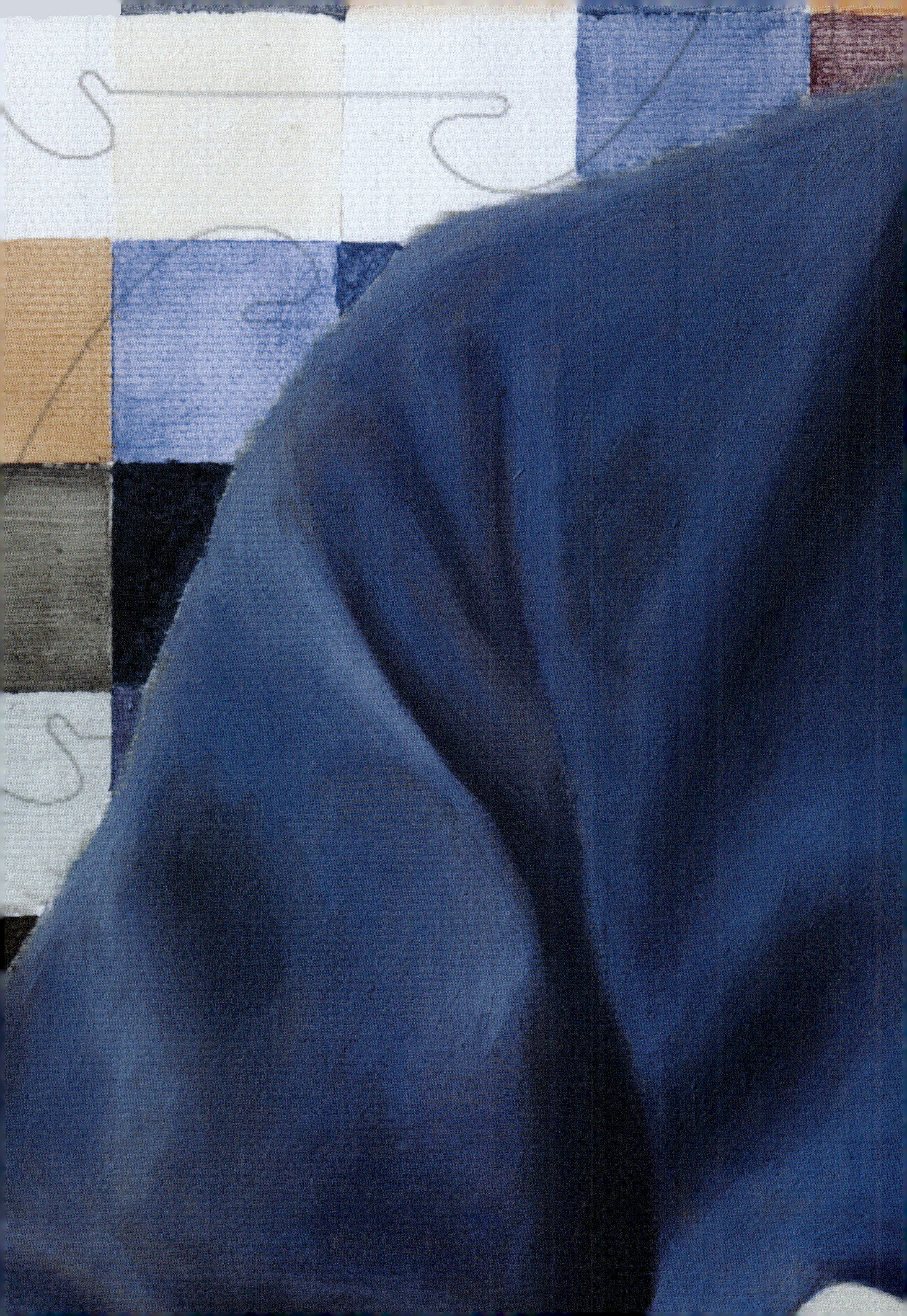

Live Study

Plates

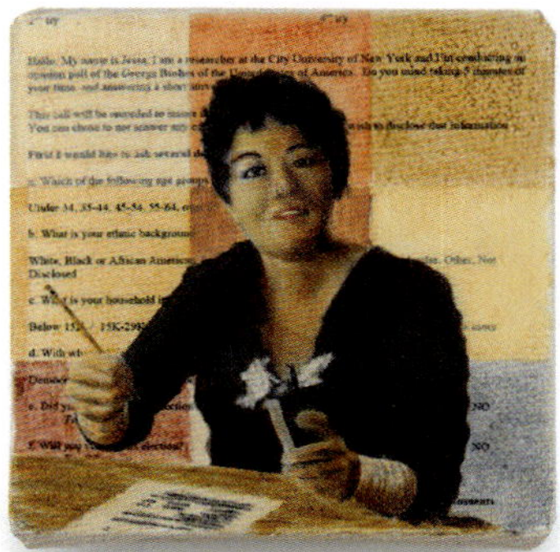

***Jessa Mendez (9 hours, 2004),* 2019**

Oil and inkjet on canvas
3h x 3w in

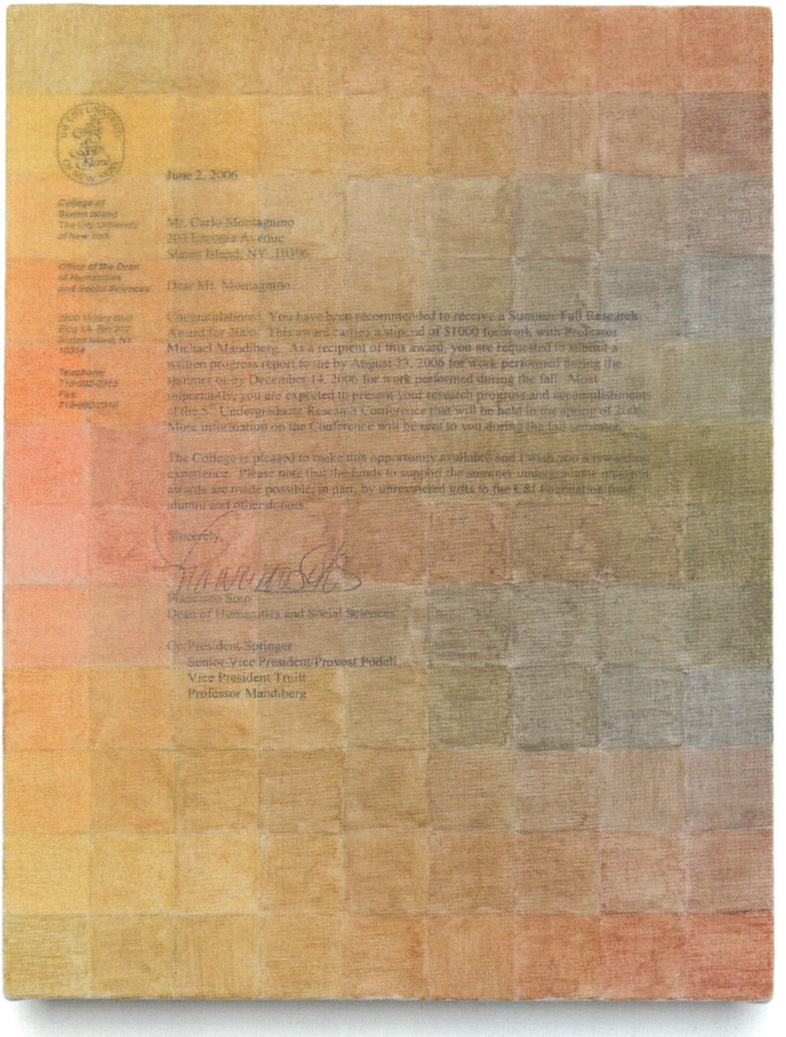

Congratulations! (Zorn vs Transparent Tint Palette), 2021

Oil and inkjet on canvas
12h x 9w in

Carlo Montagnino (502 hours, 2006-2008),
2019

Oil and inkjet on canvas
24h x 18w in

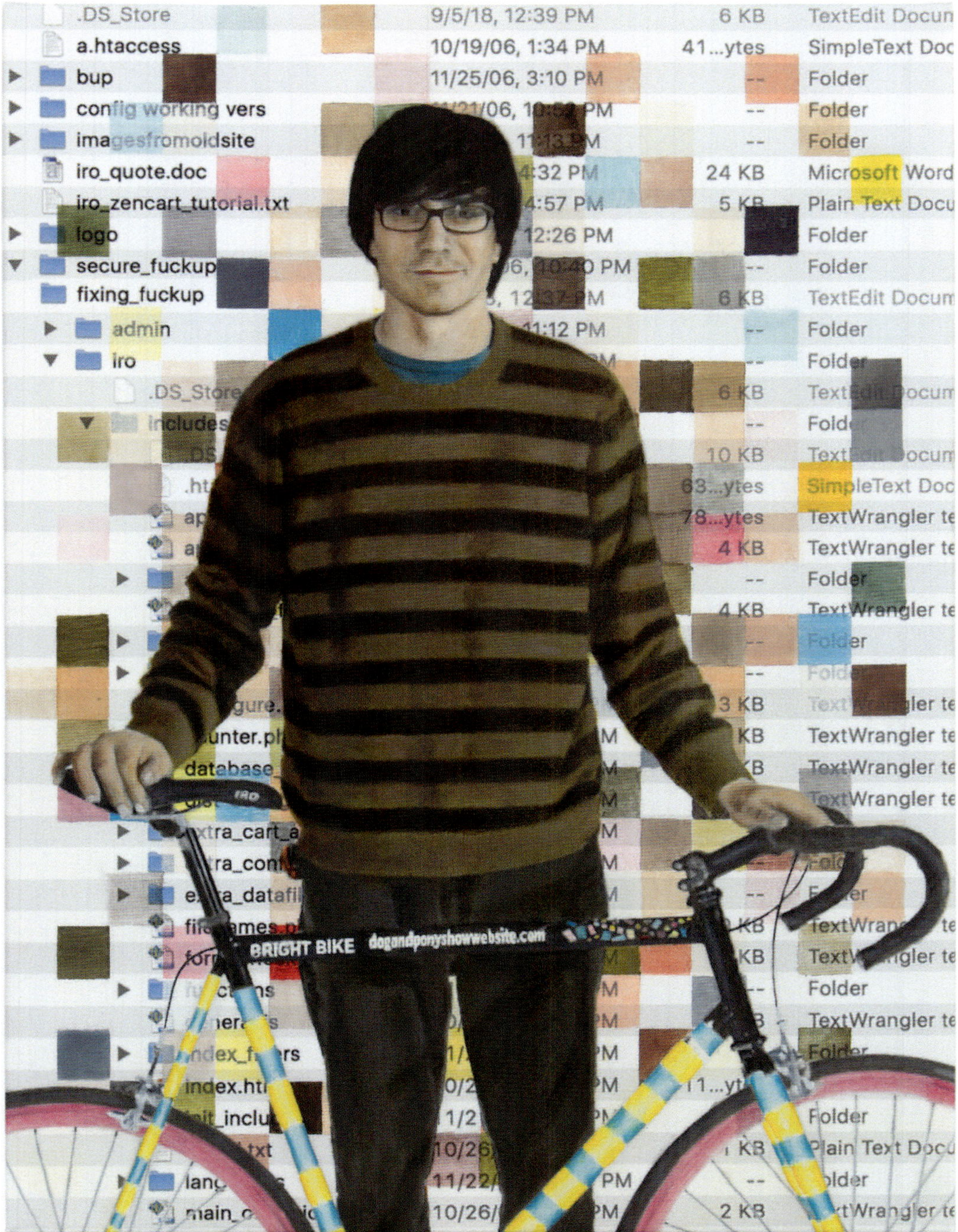

.DS_Store 9/5/18, 12:39 PM 6 KB TextEdit Docun
a.htaccess 10/19/06, 1:34 PM 41...ytes SimpleText Doc
bup 11/25/06, 3:10 PM -- Folder
config working vers Folder
imagesfromoldsite Folder
iro_quote.doc 24 KB Microsoft Word
iro_zencart_tutorial.txt Plain Text Docu
logo 12:26 PM Folder
secure_fuckup Folder
fixing_fuckup TextEdit Docum
admin 11:12 PM Folder
iro Folder
.DS_Store 6 KB TextEdit Docum
includes Folder
10 KB
63...ytes SimpleText Doc
78...ytes TextWrangler te
4 KB TextWrangler te
Folder
4 KB TextWrangler te
Folder
3 KB
TextWrangler te
TextWrangler te
TextWrangler te
Folder
TextWrangler te
Folder
Plain Text Docu
2 KB
BRIGHT BIKE
dogandponyshowwebsite.com

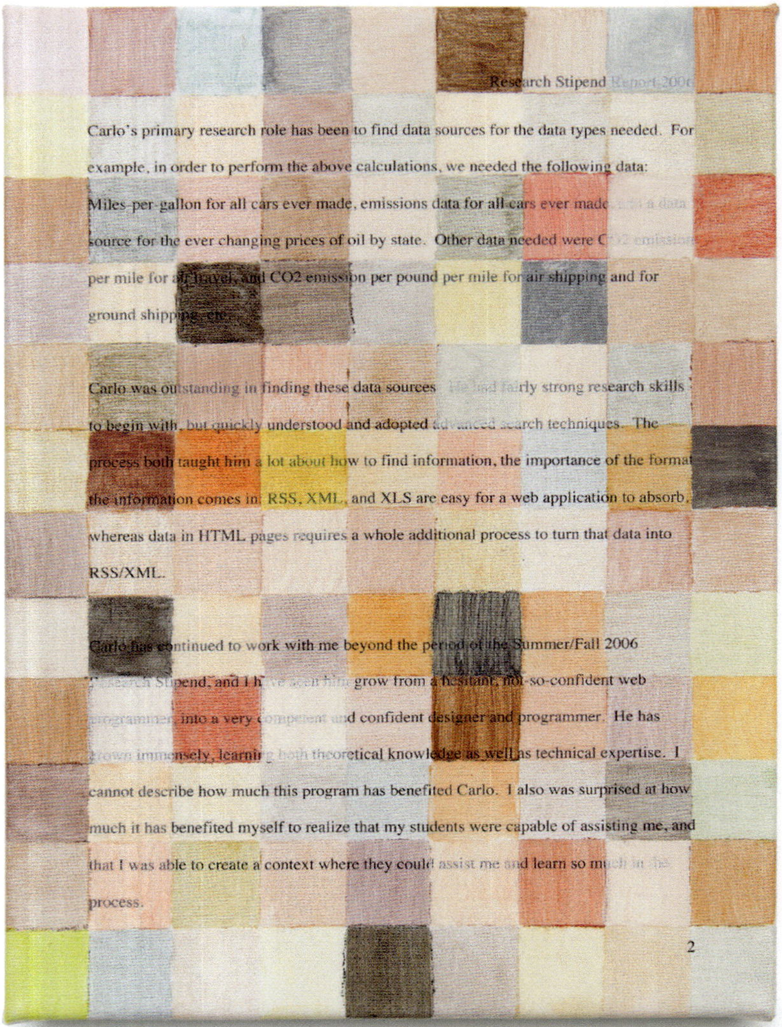

Carlo has grown immensely (Gamblin medium tests), **2018**

Oil and inkjet on canvas
12h x 9w in

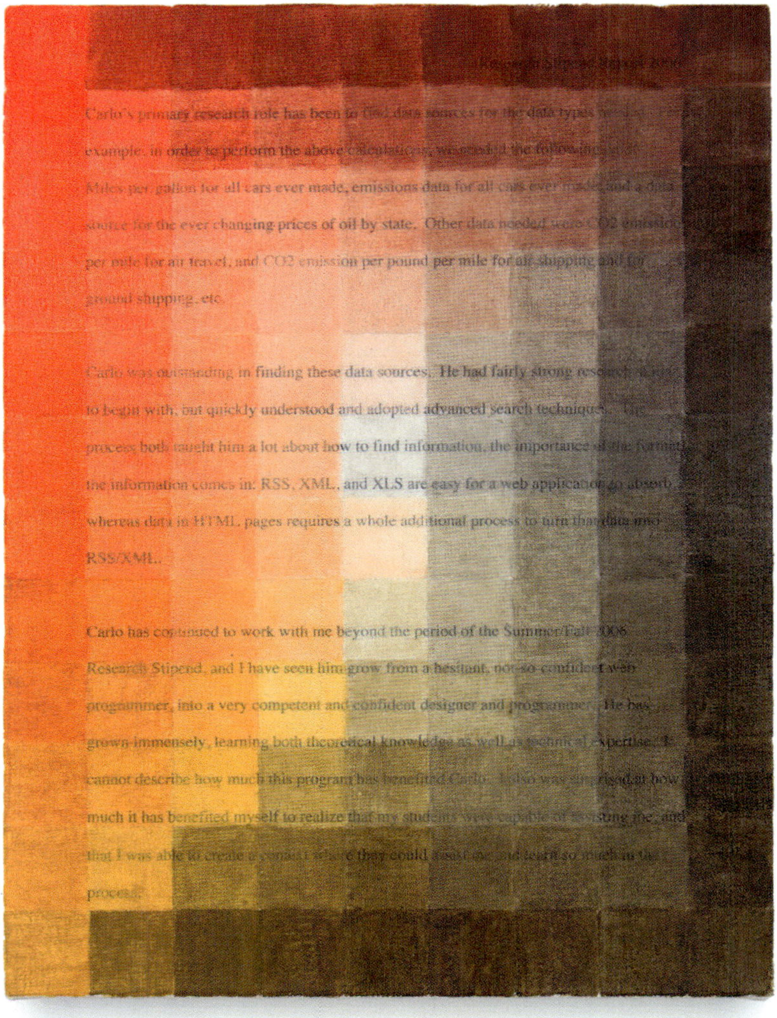

Carlo has grown immensely (Zorn Palette),
2020

Oil and inkjet on canvas
12h x 9w in

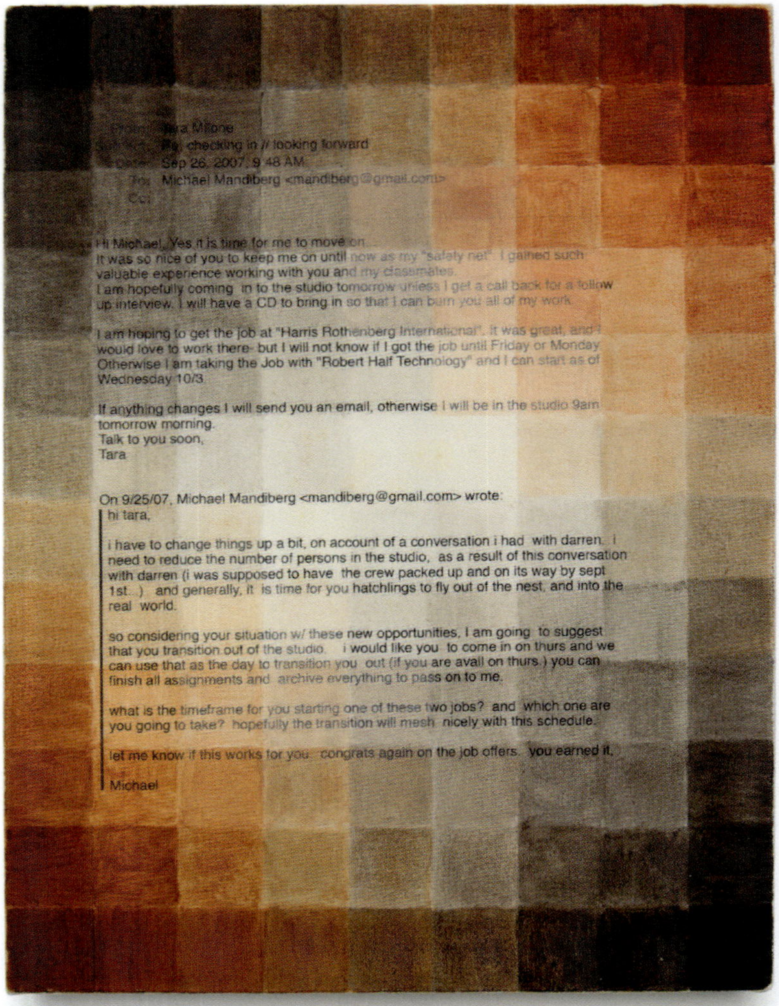

***I gained such valuable experience working with you (Bronze over Greyscale)*, 2020**

Oil and inkjet on canvas
12h x 9w in

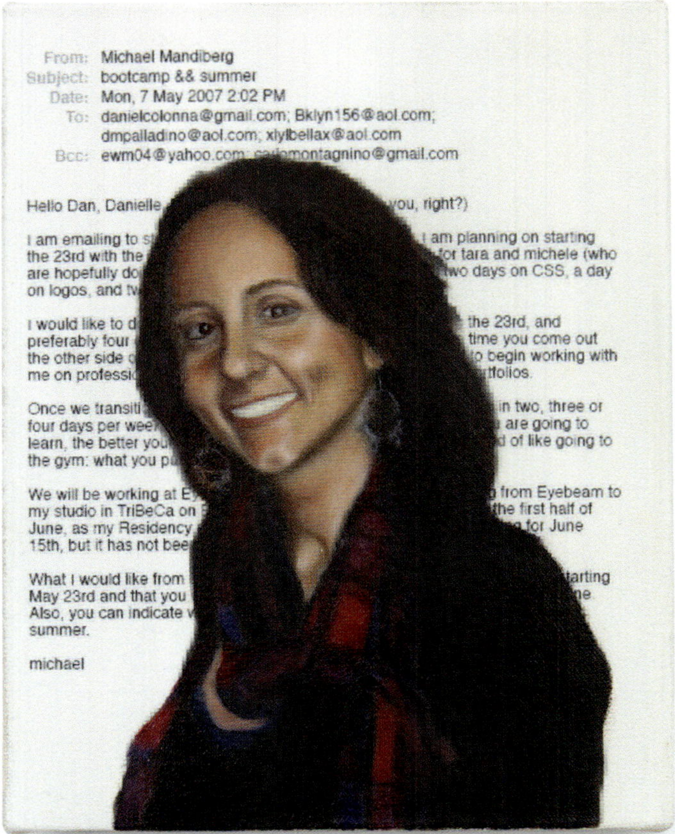

Tara Milone (59 hours, 2007), 2020

Oil and inkjet on canvas
9h x 7w in

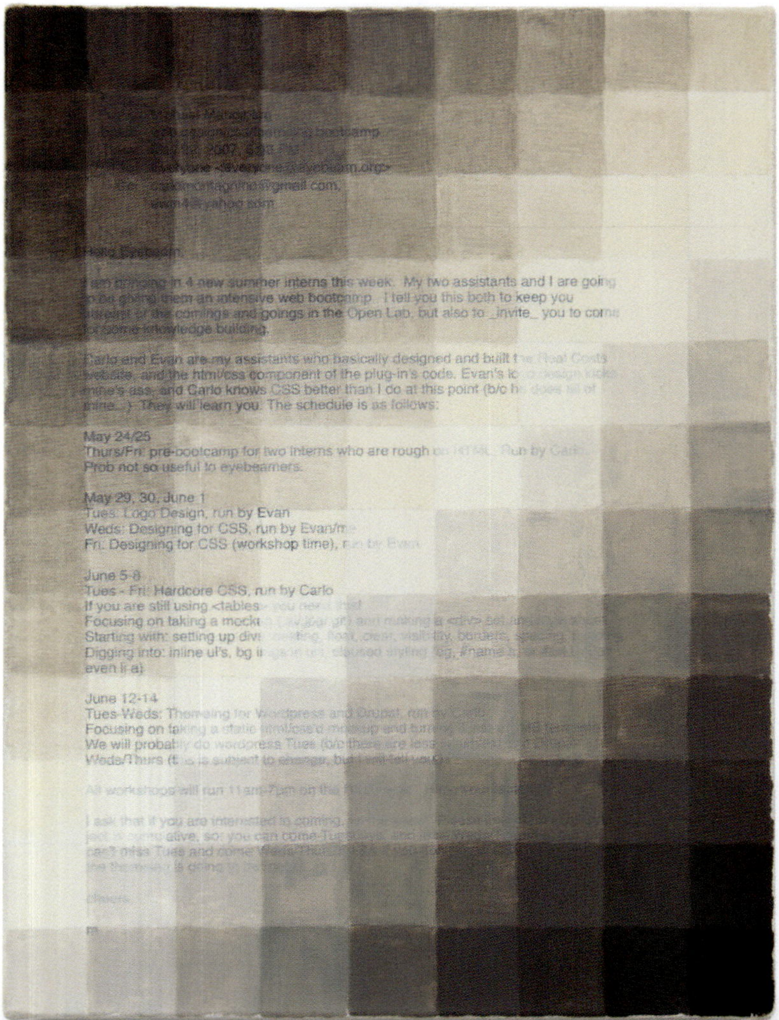

Evan's logo design kicks mine's ass (Zinc White vs Titanium White), 2020

Oil and inkjet on canvas
12h x 9w in

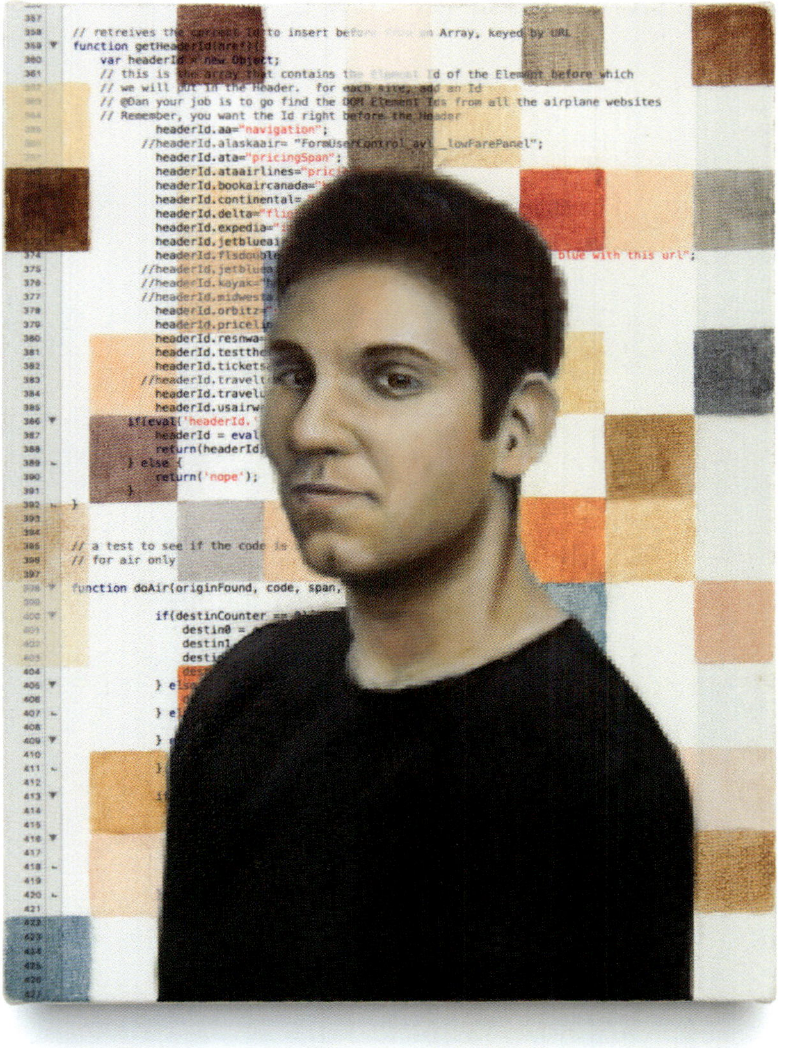

Dan Colonna (120 hours, 2007), 2020

Oil and inkjet on canvas
12h x 9w in

***Simon Jolly (636 hours, 2007-2009)*,**
2019

Oil and inkjet on canvas
32h x 20w in

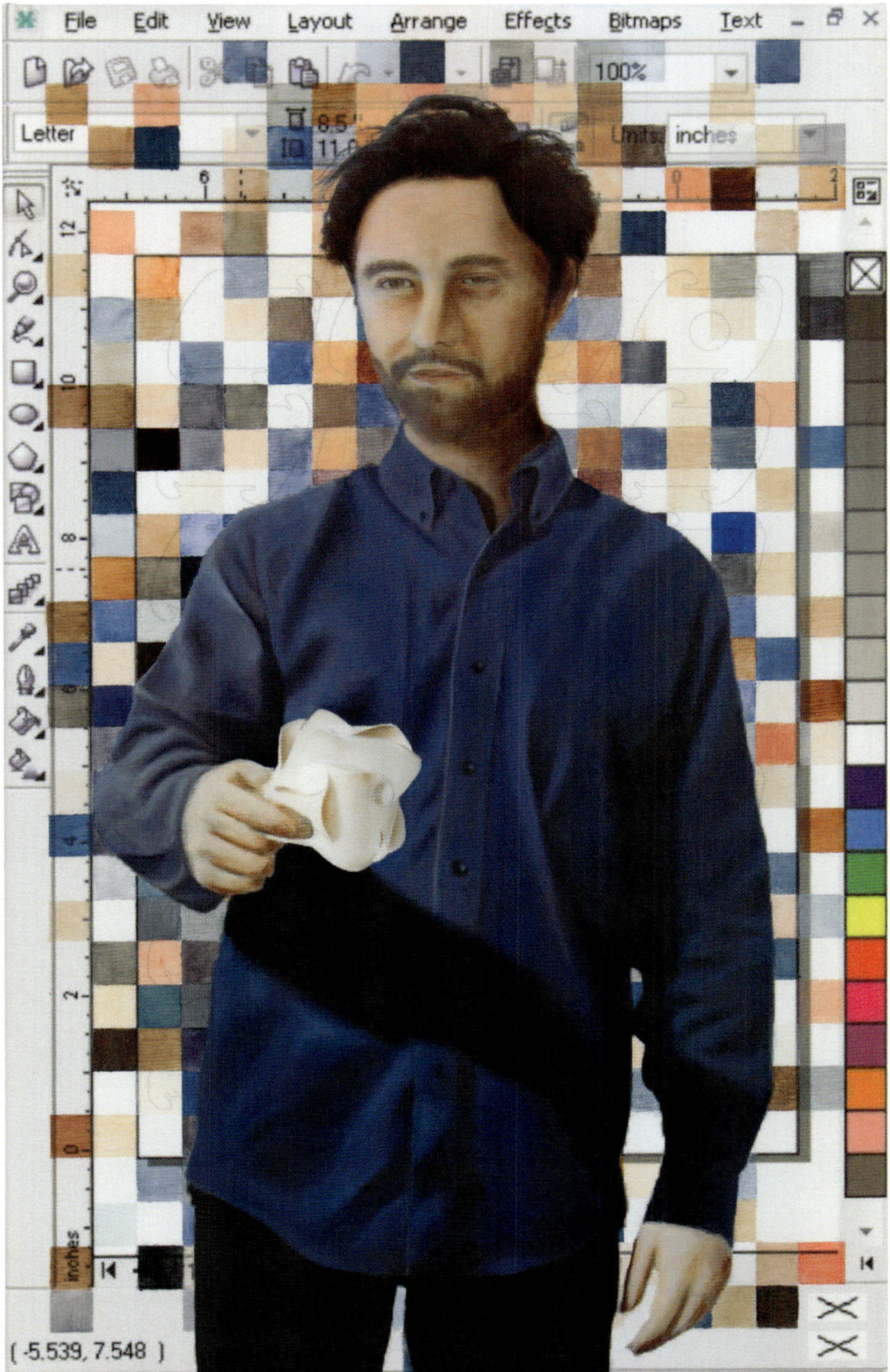
File
Edit
View
Layout
Arrange
Effects
Bitmaps
Text
100%
Letter
Units
inches
inches
(-5.539, 7.548)

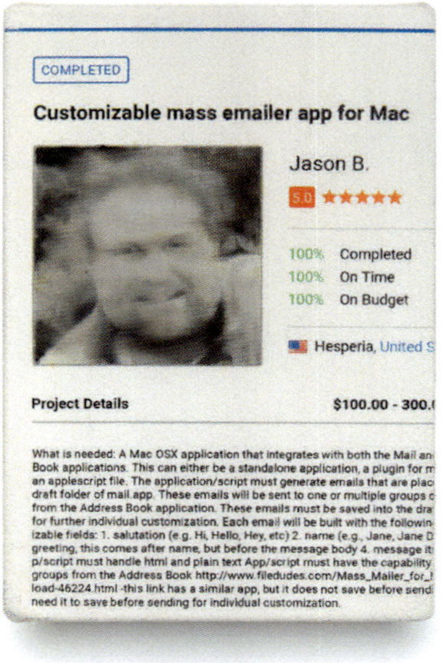

Jason B. (22 hours, 2008), 2019

Oil and inkjet on canvas
6h x 4w in

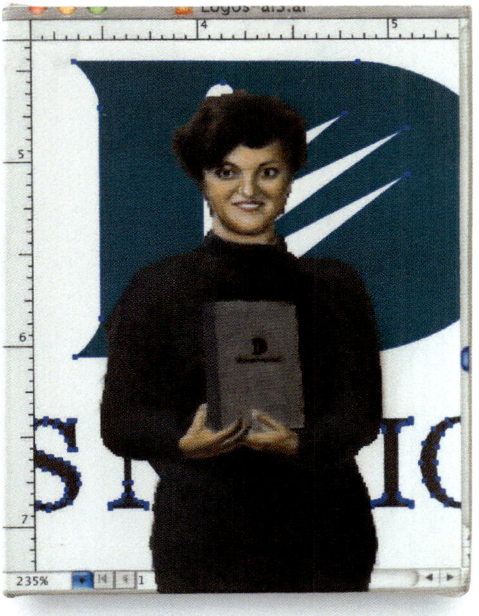

Katya Grokhovsky (48 hours, 2008), 2021

Acrylic and inkjet on canvas
8h x 6w in

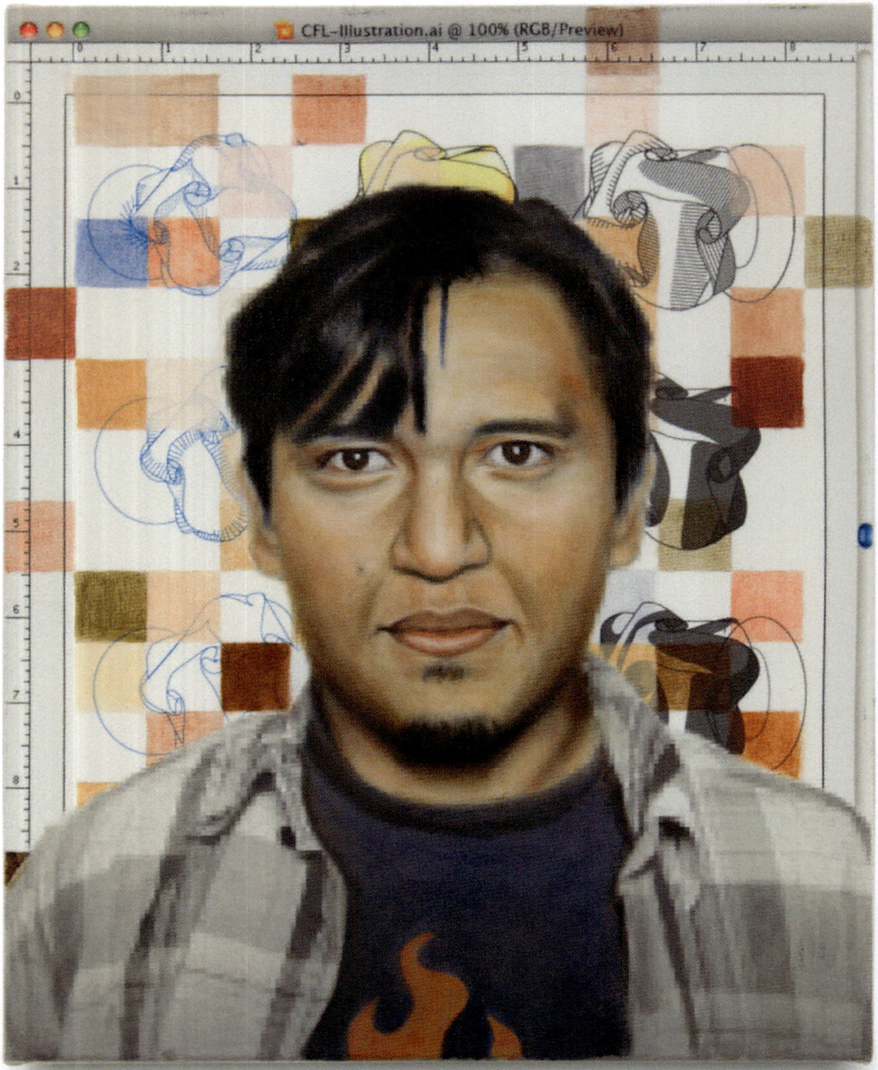

Oscar Torres (180 hours, 2008), 2020

Oil and inkjet on canvas
15h x 12w in

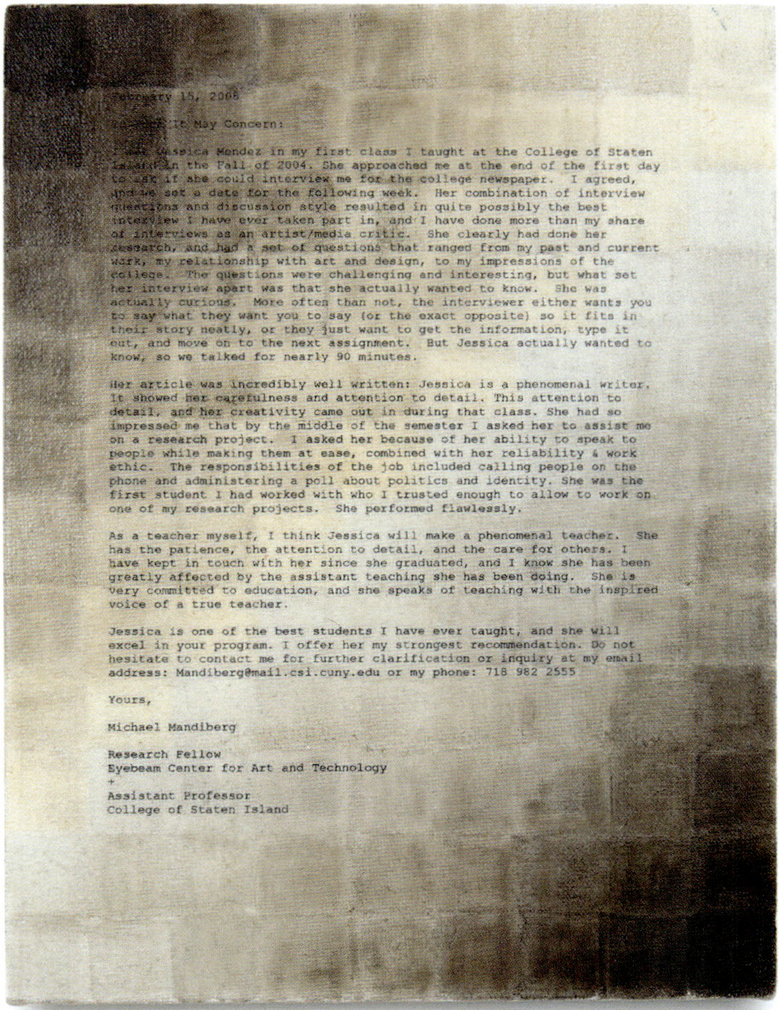

February 15, 2008

To Whom It May Concern:

I met Jessica Mendez in my first class I taught at the College of Staten Island in the Fall of 2004. She approached me at the end of the first day to ask if she could interview me for the college newspaper. I agreed, and we set a date for the following week. Her combination of interview questions and discussion style resulted in quite possibly the best interview I have ever taken part in, and I have done more than my share of interviews as an artist/media critic. She clearly had done her research, and had a set of questions that ranged from my past and current work, my relationship with art and design, to my impressions of the college. The questions were challenging and interesting, but what set her interview apart was that she actually wanted to know. She was actually curious. More often than not, the interviewer either wants you to say what they want you to say (or the exact opposite) so it fits in their story neatly, or they just want to get the information, type it out, and move on to the next assignment. But Jessica actually wanted to know, so we talked for nearly 90 minutes.

Her article was incredibly well written: Jessica is a phenomenal writer. It showed her carefulness and attention to detail. This attention to detail, and her creativity came out in during that class. She had so impressed me that by the middle of the semester I asked her to assist me on a research project. I asked her because of her ability to speak to people while making them at ease, combined with her reliability & work ethic. The responsibilities of the job included calling people on the phone and administering a poll about politics and identity. She was the first student I had worked with who I trusted enough to allow to work on one of my research projects. She performed flawlessly.

As a teacher myself, I think Jessica will make a phenomenal teacher. She has the patience, the attention to detail, and the care for others. I have kept in touch with her since she graduated, and I know she has been greatly affected by the assistant teaching she has been doing. She is very committed to education, and she speaks of teaching with the inspired voice of a true teacher.

Jessica is one of the best students I have ever taught, and she will excel in your program. I offer her my strongest recommendation. Do not hesitate to contact me for further clarification or inquiry at my email address: Mandiberg@mail.csi.cuny.edu or my phone: 718 982 2555

Yours,

Michael Mandiberg

Research Fellow
Eyebeam Center for Art and Technology
+
Assistant Professor
College of Staten Island

Jessa actually wanted to know (Warm/Cool glaze), 2020

Oil and inkjet on canvas
12h x 9w in

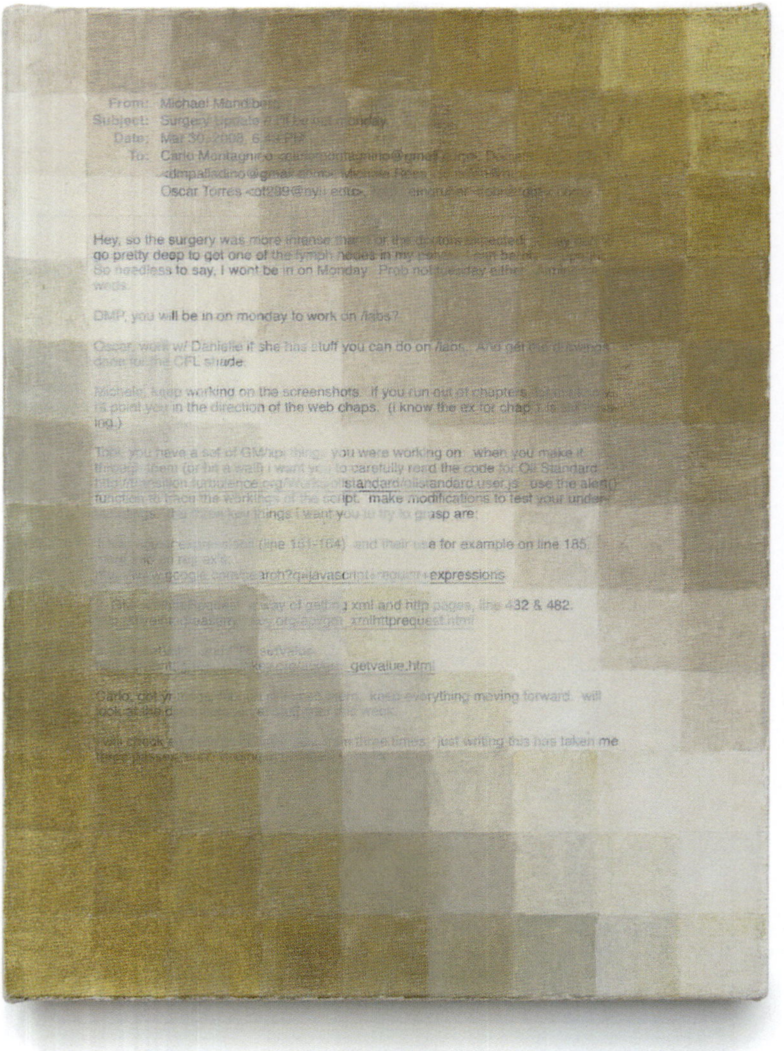

The surgery was more intense than I or the doctors expected (Zinc White vs Titanium White), 2020

Oil and inkjet on canvas
12h x 9w in

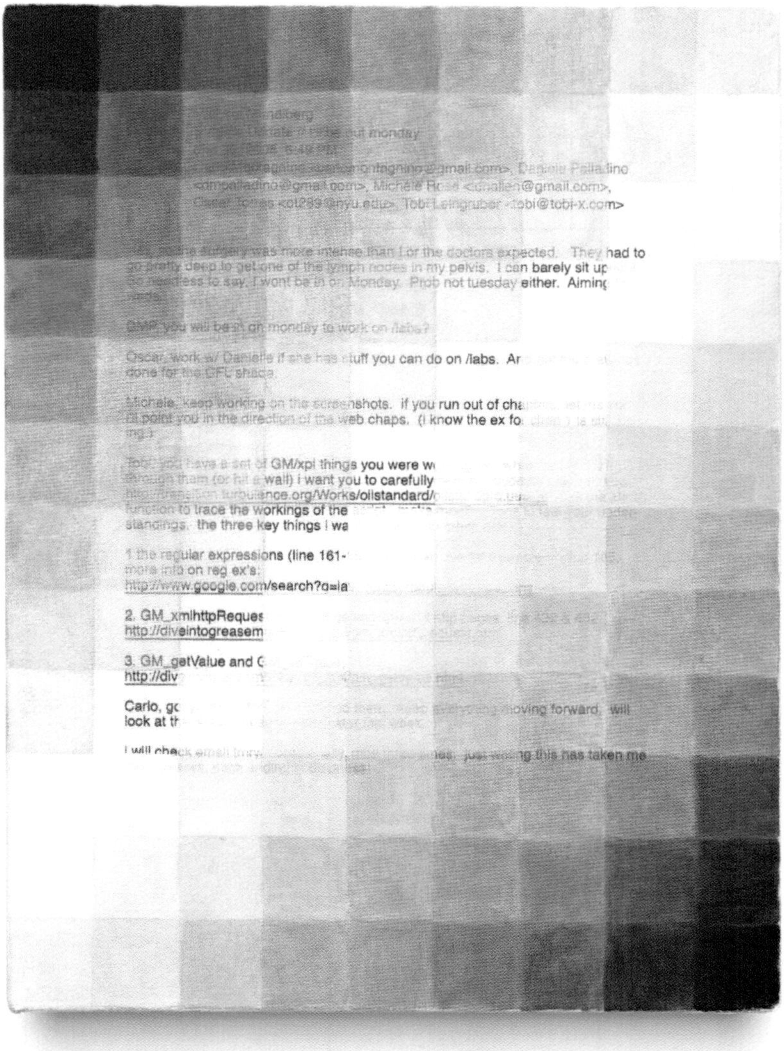

The surgery was more intense than I or the doctors expected (Transparent White vs Titanium White), 2021

Oil and inkjet on canvas
12h x 9w in

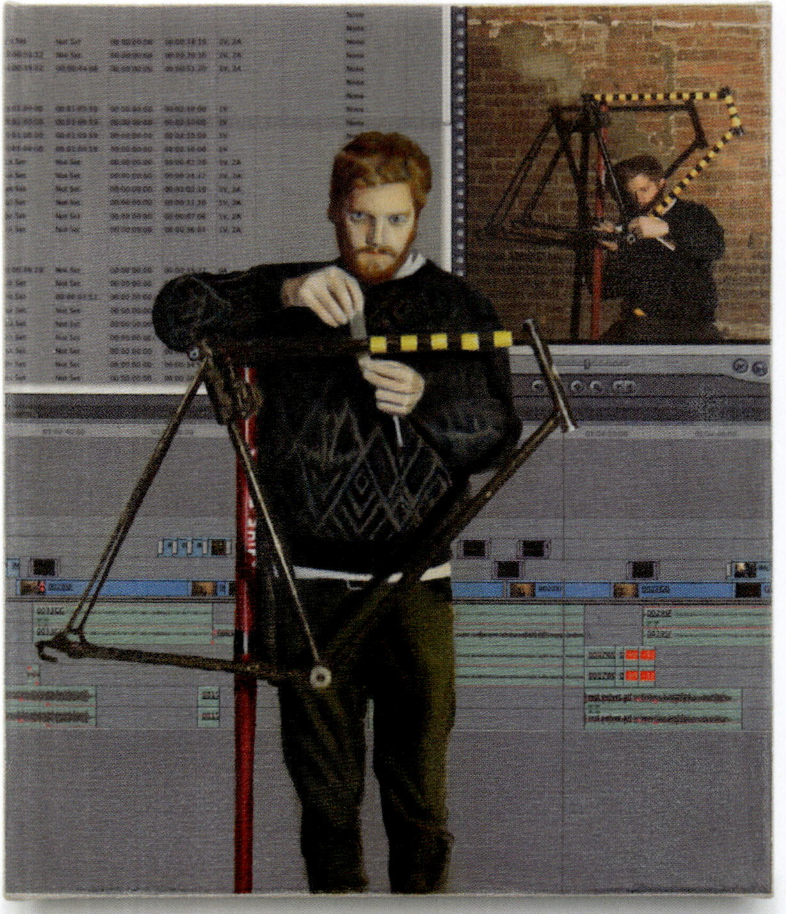

Bennett Williamson (113 hours, 2009), 2019

Oil and inkjet on canvas
12h x 10w in

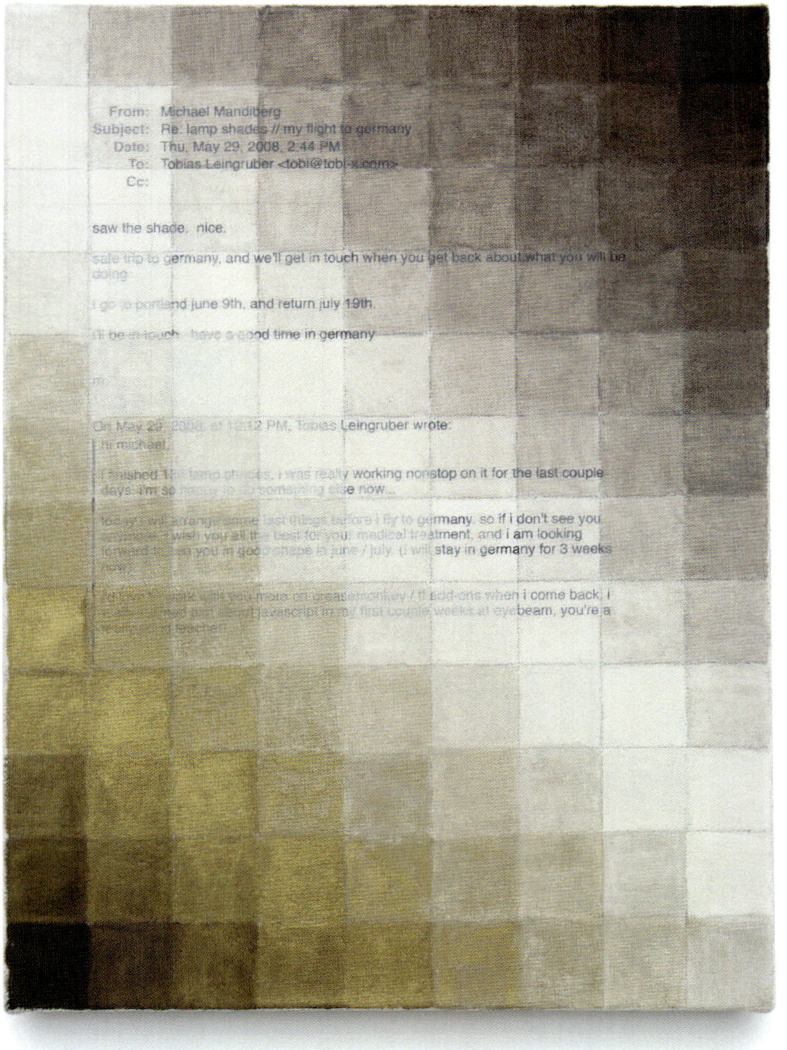

I wish you all the best for your medical treatment (Gold into Black), 2020

Oil and inkjet on canvas
12h x 9w in

***Patrick Davison (737 hours, 2008-2010)*, 2021**

Oil and UV print on canvas
31h x 24w in

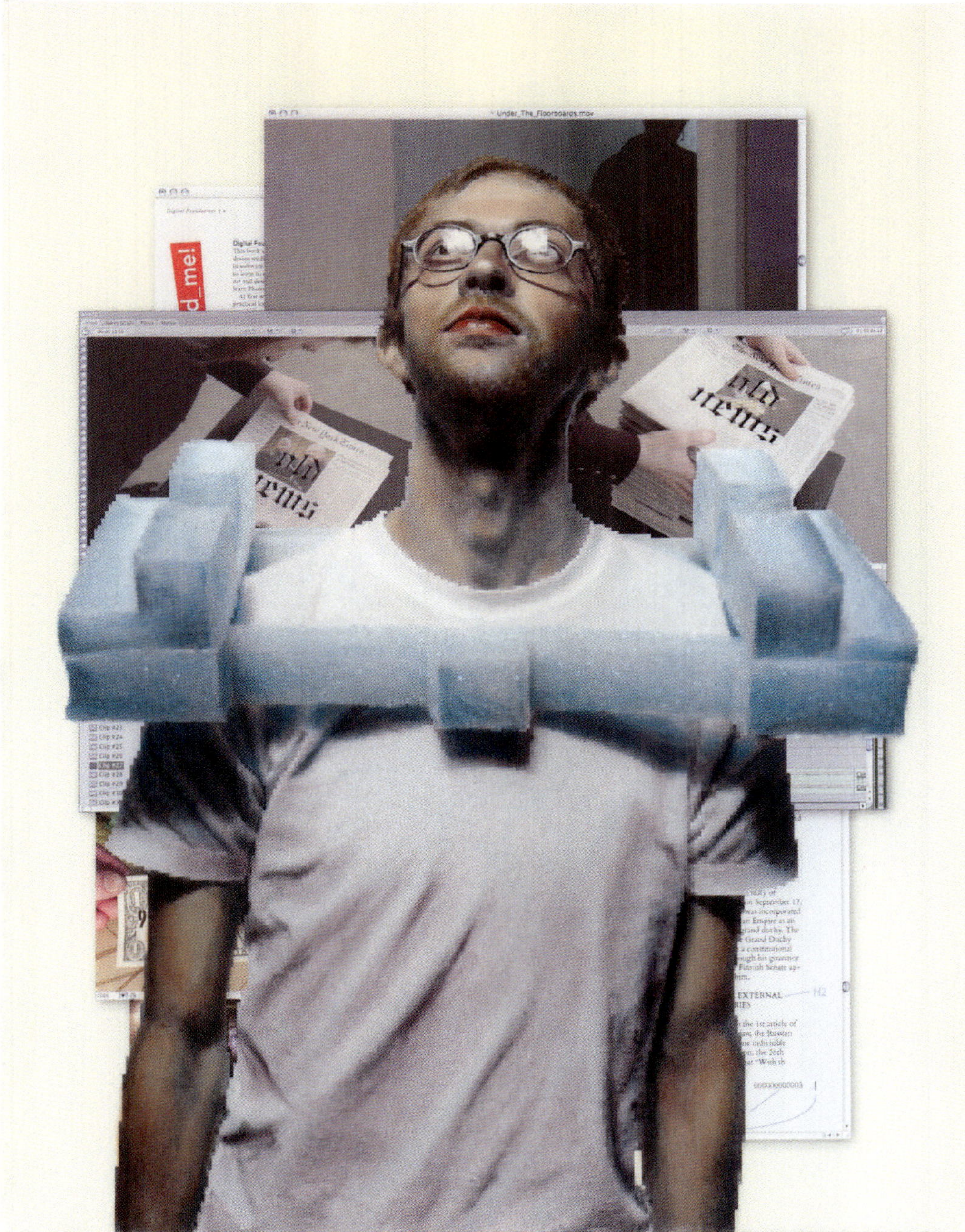

Under_The_Floorboards.mov
d_me!
EXTERNAL
H2

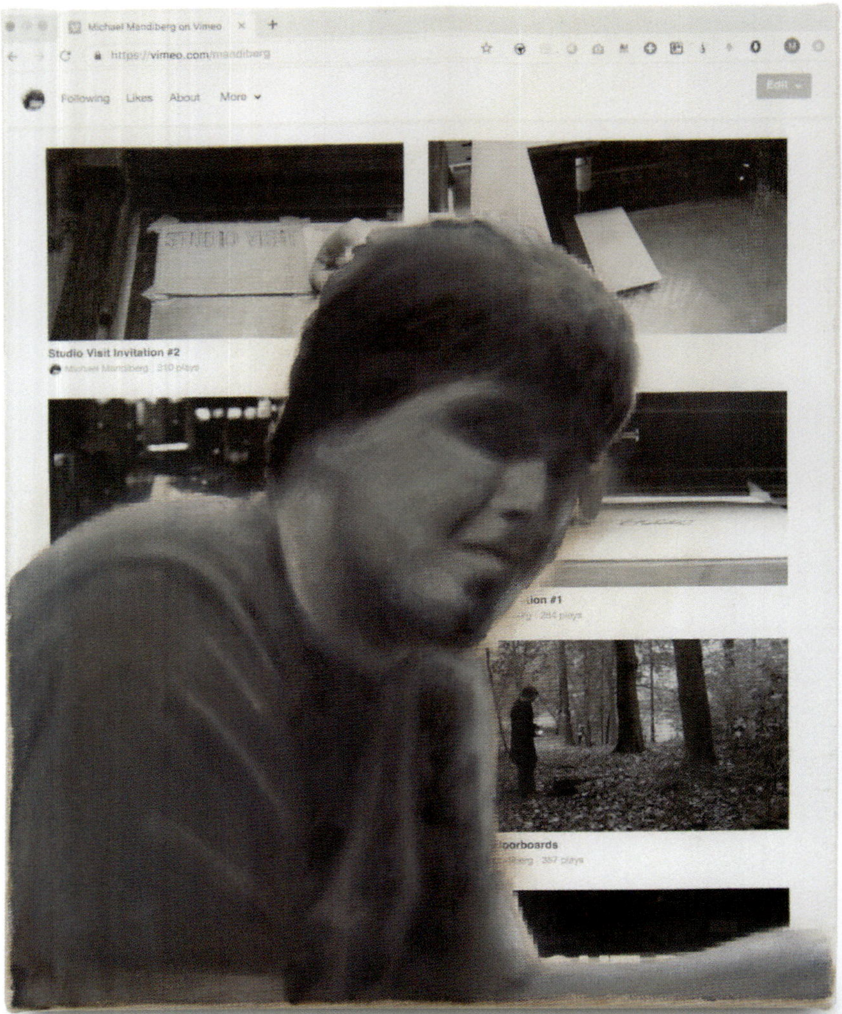

Jonathan Perlman (180 hours, 2010), 2020

Oil and inkjet on canvas
15h x 12w in

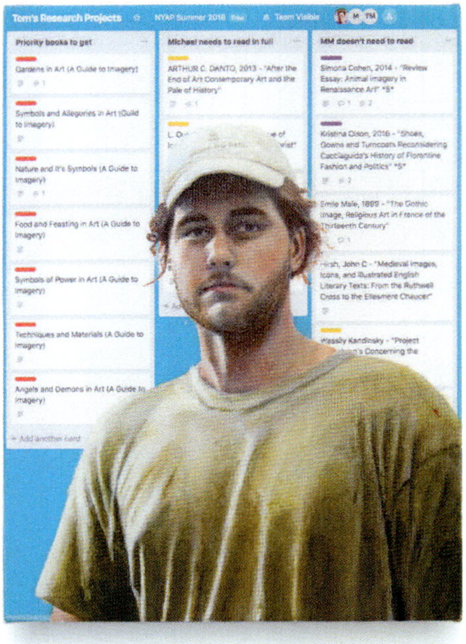

Tom Morrison (48 hours, 2018) , 2018

Acrylic and inkjet on canvas
8h x 6w in

Study for Valerie Werder, 2018

Acrylic and inkjet on canvas, mounted on panel
16h x 12w in

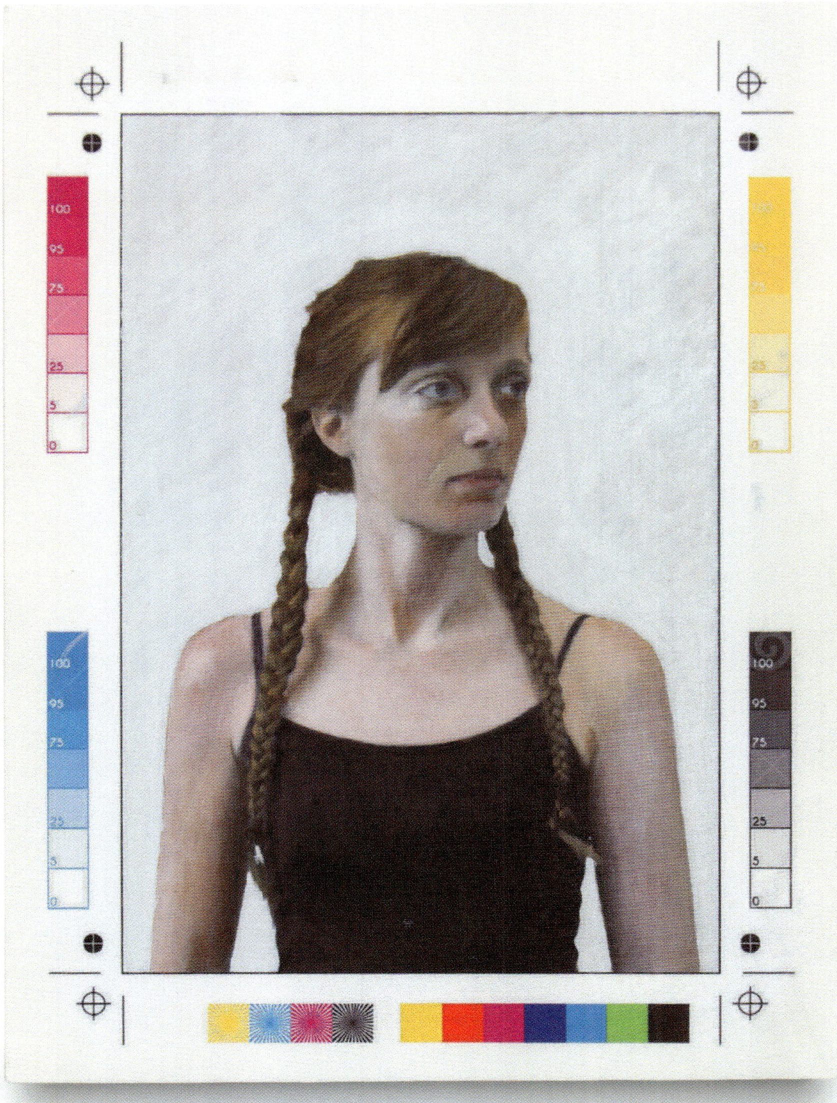
100
95
75
25
5
0
100
95
75
25
5
0
100
95
75
25
5
0

Live Study — Performing the Artist Studio

Christiane Paul

For centuries artists have represented themselves and their studio practice in their work. In his iconic painting *Las Meninas* (1656), Diego Velázquez portrays himself working on a large canvas, surrounded by several figures from the Spanish court and seemingly looking beyond the pictorial space, establishing a connection with the viewer. In *L'Atelier du peintre (The Painter's Studio: A real allegory summing up seven years of my artistic and moral life*, 1855), Gustave Courbet traced the influences on his artistic life through the representation of friends and associates, academic art practice, and members of different classes of society. Socio-political influences aside, artist studios have a long tradition as "schools" that have trained apprentices to produce the respective artist's work. Michael Mandiberg's *Live Study* both acknowledges these century-old art-historical lineages of artists' reflections on their practice and both refigures and questions them from a contemporary perspective.

At its core, *Live Study* is a durational performance: an archive of — at the time of this writing — over 800 hours of a live streamed video of the artist painting a form of self-portrait in their studio. This portrait, however, breaks with traditions in that it consists of the ongoing painterly depictions of the more than sixty assistants, apprentices, and consultants the artist has employed over the course of more than fifteen years, foregrounding labor relations rather than the figure of the artist as the focal point of creation and ingenious identity. The archive of *Live Study*, commissioned for and accessible through artport (the Whitney Museum's portal to Internet art) unfolds its meaning on several, parallel levels: it reveals the fusion of materialities that has become increasingly prevalent in contemporary art; it raises questions about the economy

Facing page:
Detail

Tobi Leingruber (100 hours, 2008-2009), 2021

Oil and inkjet on canvas
12h x 9w in

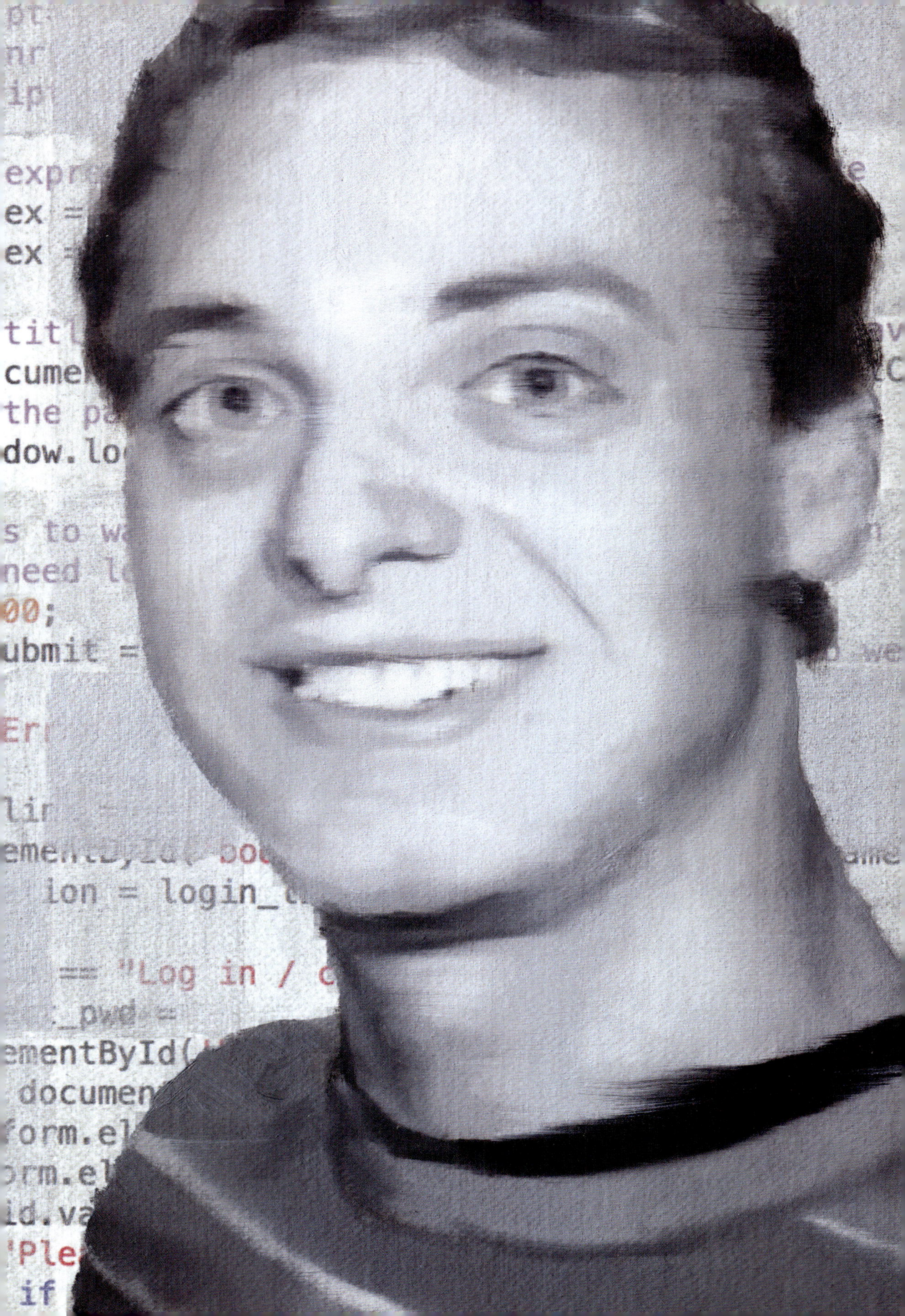

of artistic practice by memorializing the time and energy that the portraits' subjects have given to the artist; and it reflects on some of the complexities of learning, support, and collaboration.

Live Study comprises layers of encoded materialities. On the purely technological level of broadcasting, the piece consists of an encoded video transmission, but Mandiberg also developed a system for inscribing layers of meaning into the project. Each canvas is sized according to the total hours worked by the respective person at one square inch per hour, with canvas sizes ranging from a 4 × 5 inch miniature of a technical consultant to a 4 × 5 foot portrait of a long-term assistant. The background of each painting is often a 1 × 1 grid of semi-transparent glazes of all the colors used in that painting. The grid not only calls attention to the hours of labor represented by the scale of the piece and alludes to minimalist painting, but also references key elements of digital culture and practice — the Photoshop transparency grid and pixelation. Interface elements of word processing and project management software also appear in the background of the portraits. Michael Mandiberg starts the process of creating the paintings by compositing a grayscale digital collage from photographs of the respective subject from the time they worked in the studio. The artist then prints that image on a specially prepared canvas and layers oil paint in gray monochromes, and translucent glazes creating the impression of coats of paint glowing through one another. In its fusion of online broadcasting, the digital tool kit, and traditional painting, *Live Study's* confluence and convergence of digital technologies in various materialities captures the contemporary artistic practice that is often described as post-digital. *Live Study* functions as an evolving virtual document of how digital influences become embedded in the creation of paintings and how our relationship with materialities, such as painted portraits, has been changing.

The live transmission of the act of painting, as well as the interface and software elements embedded in the portraits, convey both the aesthetics and characteristics of digital communication and labor. Over the course of the more than 15 years that Michael Mandiberg has had assistants working in their studio, immaterial digital labor became a dominant form in work environments. Italian sociologist and philosopher Maurizio Lazzarato identified the series of activities that are not normally recognized as "work" — such as defining and fixing cultural and artistic standards, tastes, and consumer norms — as one of the key elements of immaterial labor.[1] Building on Lazzarato's ideas, theorist and activist Tiziana Terranova argues that digital culture

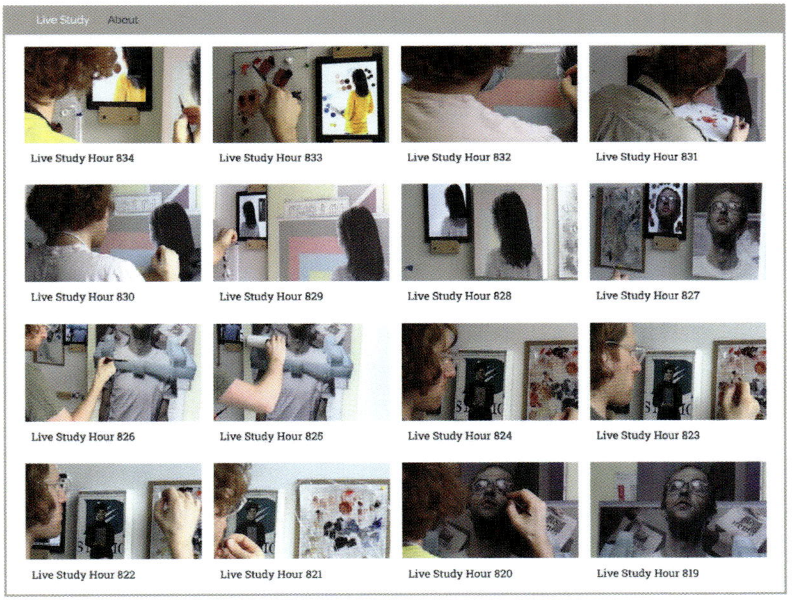

Michael Mandiberg, ***Live Study* 2018-Ongoing Durational performance,** HD video on website 834 hours, as of October, 2021

extracts value out of continuous, updateable work and that commodities become increasingly ephemeral and more of a process than a finished product.[2] In a subtle way, *Live Study's* depiction of the work-intensive process in which both the creator and subjects of the portraits have been engaging hints at the complexities of immaterial labor in the art world and artist studio. Mandiberg also positions the paintings within the economics of the art market by creating two versions of each portrait: one priced according to the total wages earned by the portrait's subject and entering the market; the other to be gifted to the subject, invoking the alternative model of a gift economy based on reciprocal relationships without explicit expectations of rewards.

Reciprocity is an underlying theme in *Live Study's* performative exploration of the learning process and exchange that plays a key role in artists' studios. While the work may formally evoke YouTube instructional videos that teach mastery of any given subject, it effectively functions as documentation of a learning process. With *Live Study*, Michael Mandiberg introduces painting to their practice

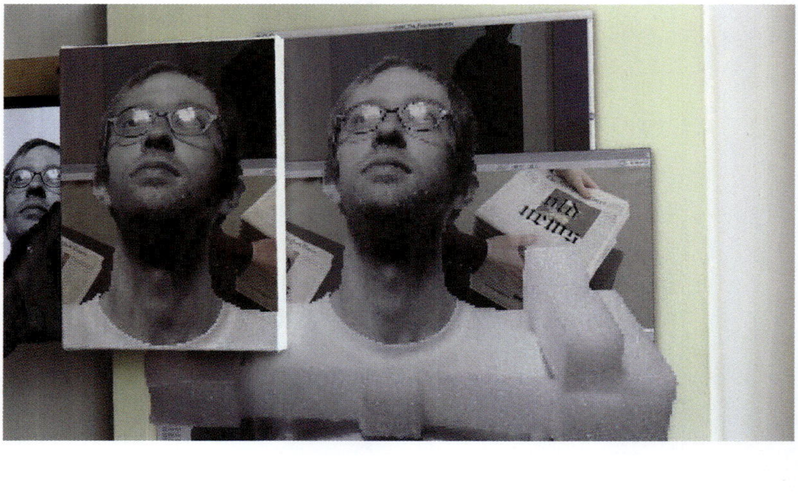

Michael Mandiberg, Still from ***Live Study 2018-Ongoing Durational performance***, HD video on website 834 hours, as of October, 2021

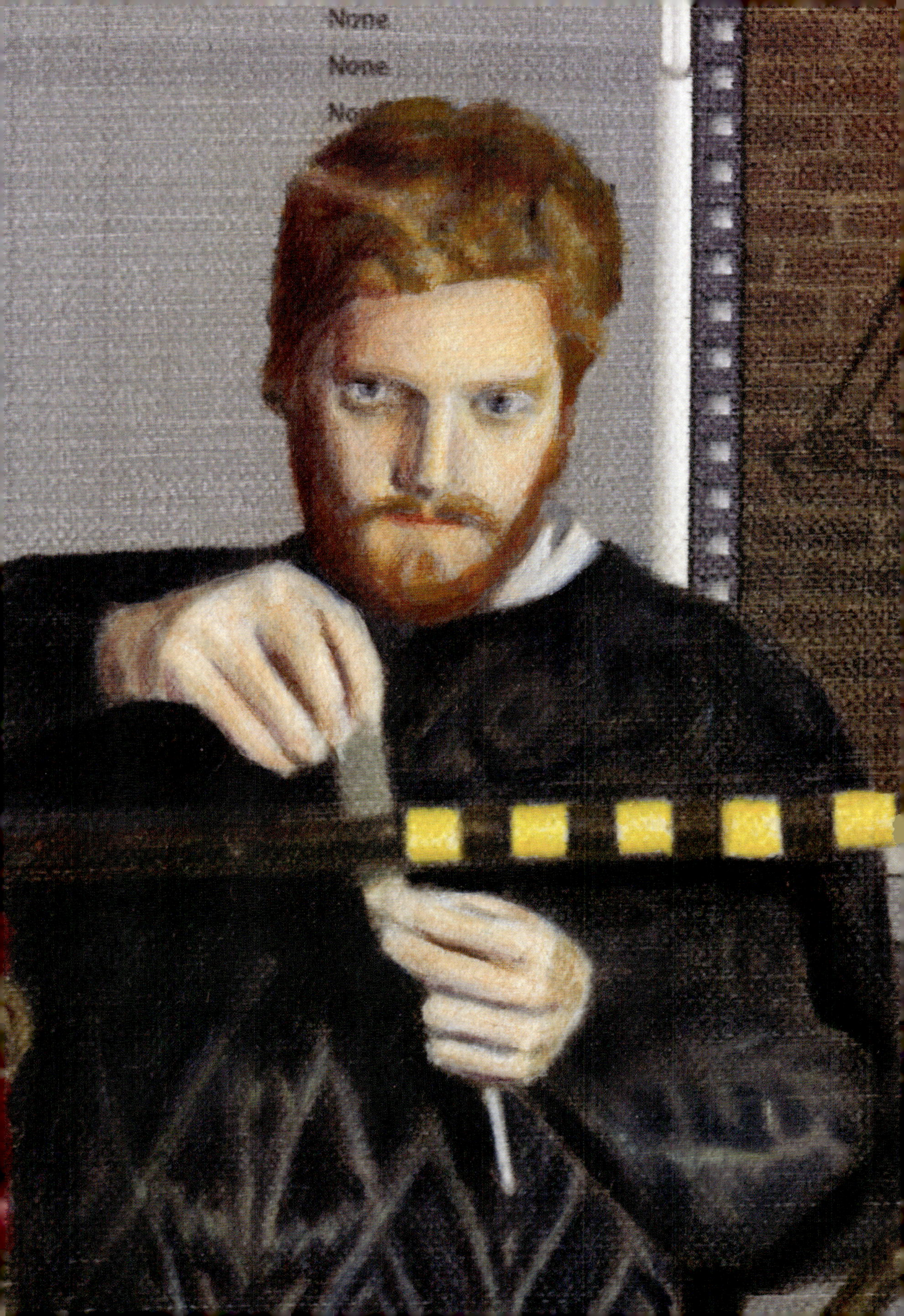
None.
None.

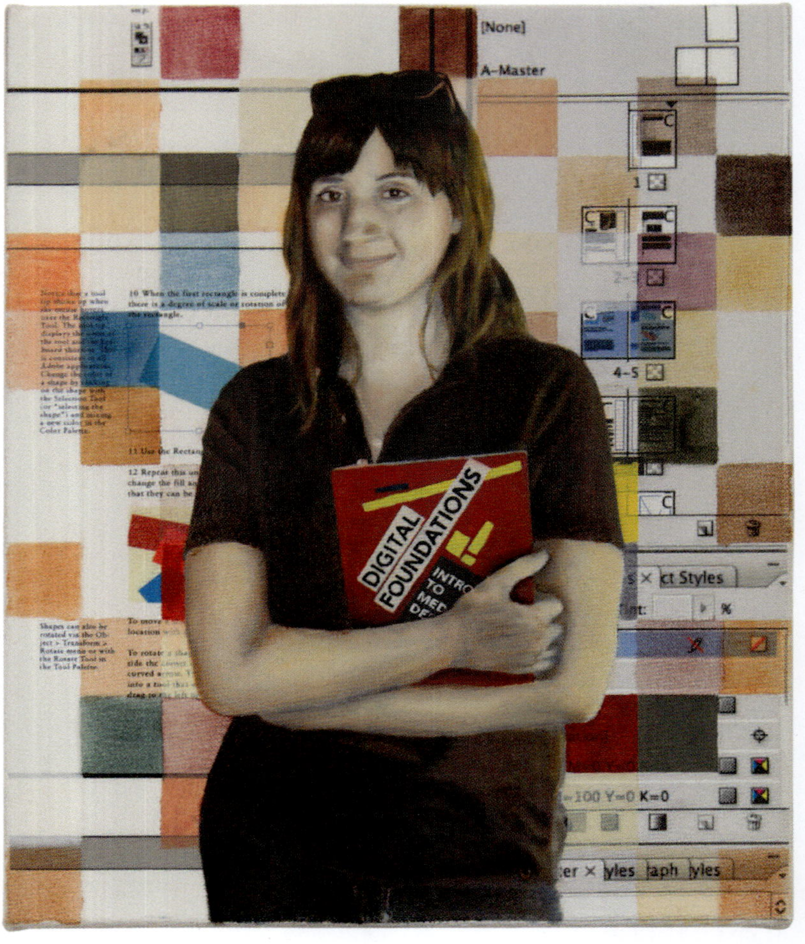

Tara Romeo (118 hours, 2008), 2019

Oil and inkjet on canvas
12h x 10w in

for the first time, and the videos become a record of the often-tedious journey of learning, with all its false starts, revisions, and profound frustrations. The archive of videos traces how the artist's skills gradually improve, echoing the ways in which Mandiberg's assistants may have built their own aptitudes while working in the studio. The concept for *Live Study* was also informed by the fact that Mandiberg's assistants temporarily became surrogates, overcoming limitations of the physical body, when the artist was unable to work while undergoing aggressive treatment for cancer that began in 2008.

Live Study is a portrait of the artist as a student of technique and manager of collaborative labor. Artistic identity can always be seen as a social and sometimes political performance, but Mandiberg underscores how its economic force often obscures labor relations. The work reveals the intricacies and instability of "the artist" as a clearly delineated creative entity and makes that identity dissolve into a socio-economic system. This theme is not new to Mandiberg's practice but has surfaced in previous pieces. Their work *Quantified Self Portrait (One Year Performance)* (2016–17), for example, is a video consisting of screenshots and photographs captured by the artist's computer and iPhone every fifteen minutes, a technique commonly used to monitor freelance workers. Rather than fulfilling the promise of constant self-tracking as a method for self-knowledge and self-improvement propagated by the Quantified Self movement and wellness industry, Mandiberg's durational performance highlighted their daily work routines and labor practices.

In the literal sense, an "endurance performance" requires integrating multiple physiological and psychological systems. It involves building physical skills while keeping mental focus in a long-term commitment, overcoming the weariness resulting from repetitive action. Michael Mandiberg's *Live Study* explores artistic practice as an endurance performance embedded in a complex system of personal and socio-economic relationships.

1. Maurizio Lazzarato, "Immaterial Labor" in Virno, Paolo and Hardt, Michael (eds.), Radical Thought in Italy: A Potential Politics (University of Minnesota Press, 1996), pp. 142–157

2. Tiziana Terranova, "Free Labor: Producing Culture for the Digital Economy" in Social Text, 63 Vol. 18 No. 2 (2000), pp. 33–58

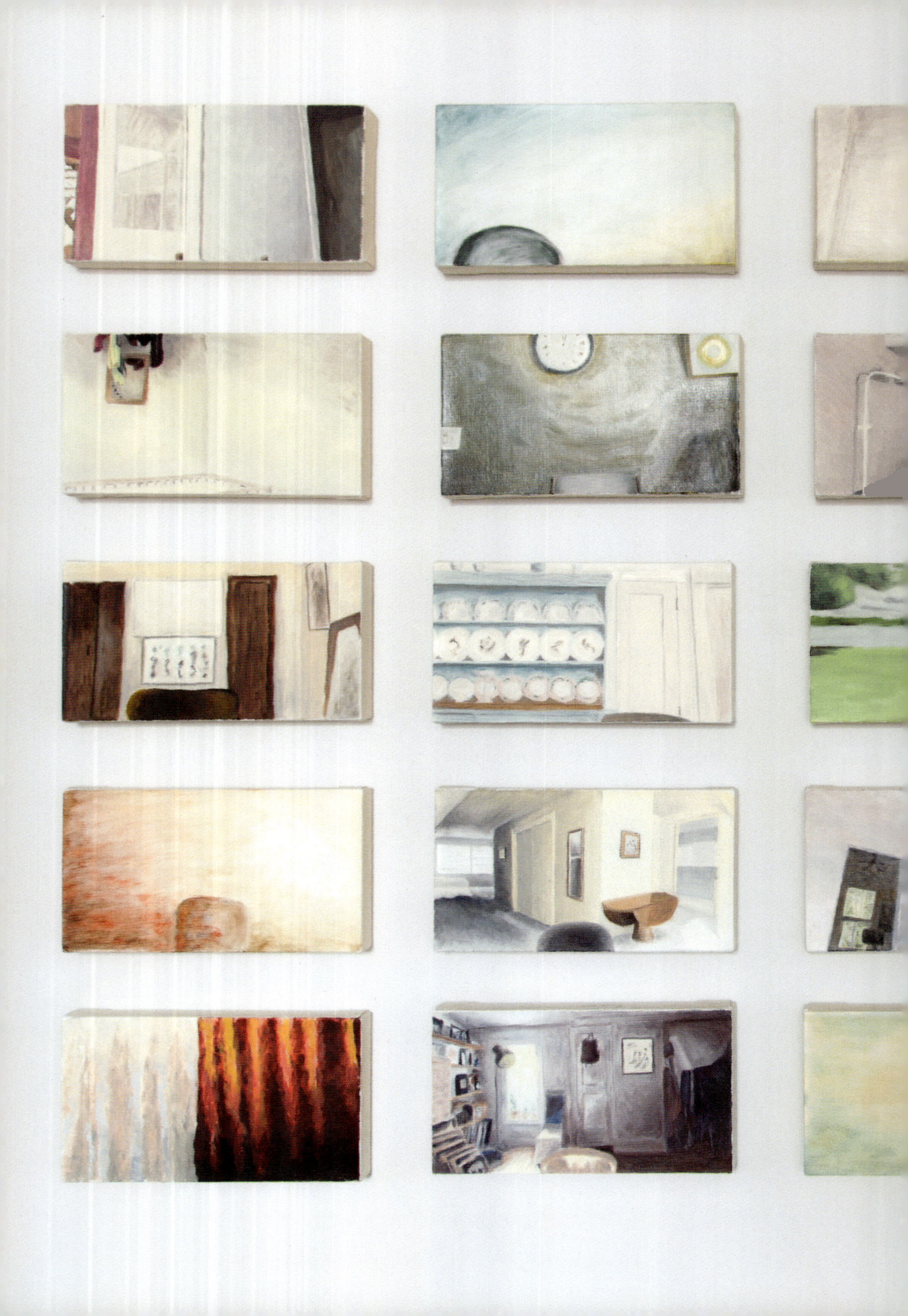

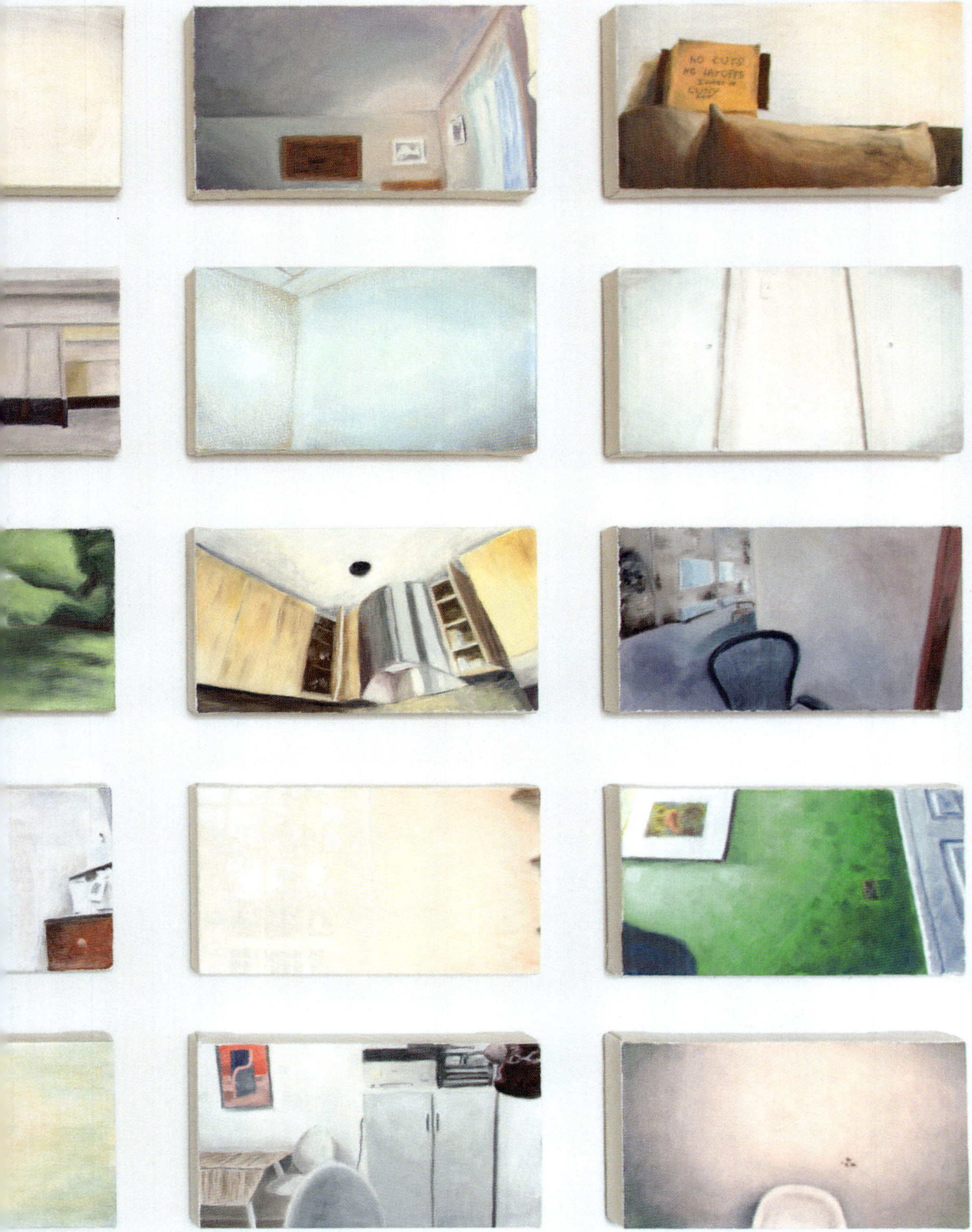
NO CUTS
NO LAYOFFS

The Zoom Paintings

Michael Mandiberg

January 21, 2021

As New York paused for COVID-19, almost all my human contact and communication ceased. Like many information workers, I found myself on hours of daily video calls. My social life became rotation of familiar faces punctuating the workday, sitting across from me in unfamiliar rooms. Unable to leave my own apartment, I have visited makeshift office spaces, the kitchens of coworkers and family members, friends' dinner tables, and the childhood bedrooms of my students.

In response, I have begun a series of paintings that capture and collect these interactions. Each painting corresponds to a video call I have made during the pandemic and depicts the room of one of the people on my call. Sized at six by eleven inches, each painting reflects the Zoom call's sixteen-to-nine proportions. I paint these rooms without the person sitting before the camera. These paintings are not about the person who inhabits the room I painted. Nor are these paintings about what the books on their bookshelves may or may not tell us about them, unlike the hundreds of pandemic articles that earnestly analyzed these bibliographic mise-en-scènes. Instead, they represent the interchangeability of these spaces. The nowhereness of the blank wall. The aloneness. The loneliness. And the dissociation. In doing so, these paintings serve as a record of interactions marked by absence.

These webcam images constitute a new way of seeing our domestic spaces. In drawing classrooms students learn to draw interior spaces with two-point perspective, always keeping the vertical lines parallel. Trained photographers know to use their camera (or Photoshop) to ensure the vertical lines are always parallel. These new disorienting extra-wide-angle images break all these rules of "good

Facing page:

Detail view of Zoom Paintings installation

Center top:
Art + Feminism, Phase 2 part 1a, 6:00 — 9:00 PM, December 15, 2020 (#43), 2020
Oil on canvas
6h x 11w in

Center bottom:
Graduate Center IT Committee, 8:00 — 9:00 PM, March 18, 2021 (#56), 2021
Oil on canvas
6h x 11w in

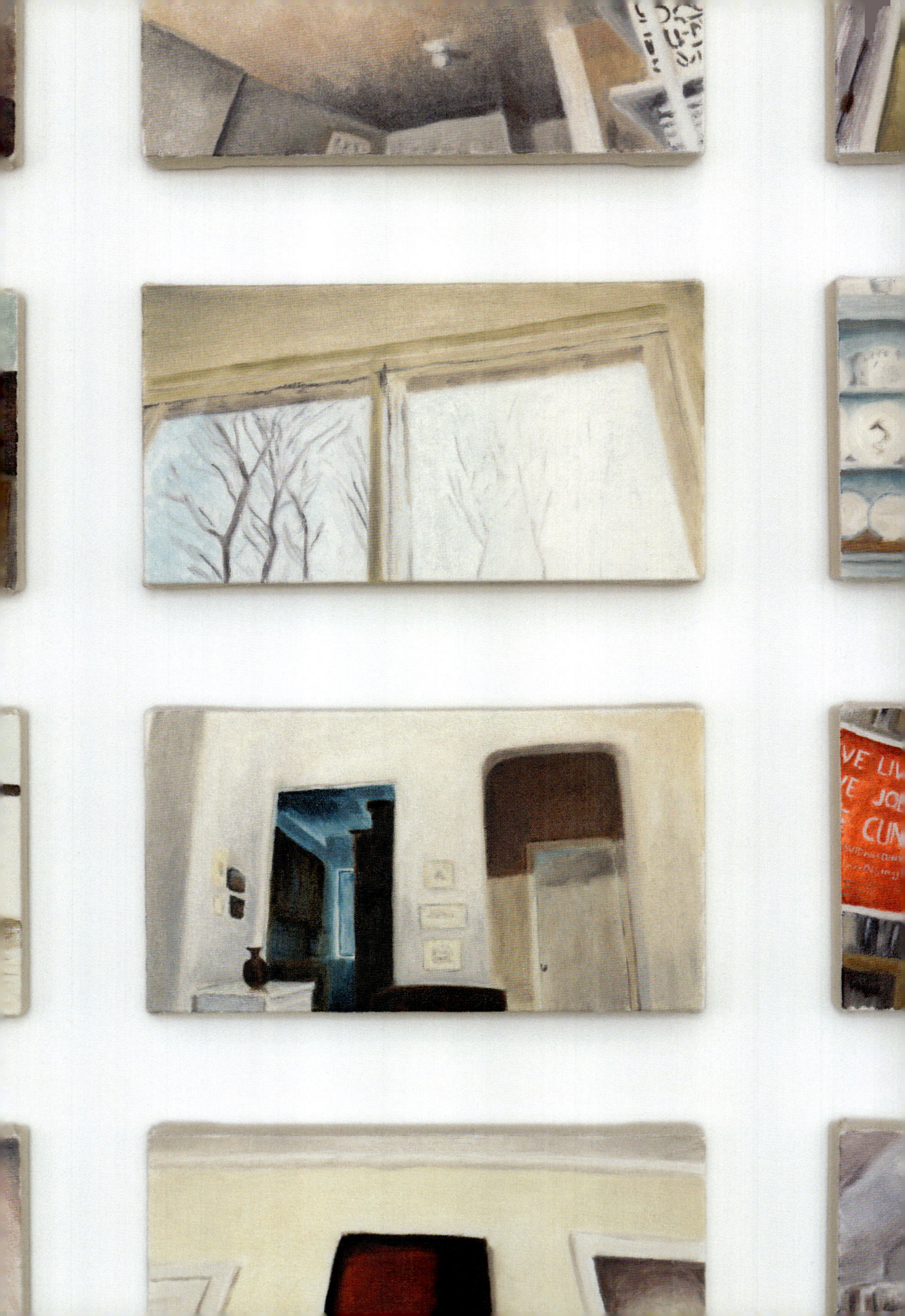

photography." They skew familiar domestic spaces, pointed up into the corners of rooms in destabilizing three-point perspectives, where not a single line runs parallel or perpendicular to the frame of reference. Film-camera family snapshots, smart phone apartment rental pix, and most other genres of "bad image" are taken at eye level, at standing height near the middle of the room, looking slightly down. In these webcam images, we almost always look up, from laps, kitchen tables, coffee tables, or the desks of improvised home offices. And sometimes, when someone puts down (or drops) their phone, we become a truly disembodied eye, laid down flat on the table facing up at the ceiling. These techniques of observing domestic space produce a kind of vision dissociated from the body. Unlike the disembodied vision of drone photography, virtual reality, or images generated by machine learning, these images remain embodied, though dissociated. They are beside the body, simultaneously present and absent.

Facing page:

Detail view of Zoom Paintings installation

Center top:
ella call, 11:00 AM — 12:00 PM, September 22, 2020 (#28), 2020
Oil on canvas
6h x 11w in

Center bottom:
ITP advisory meeting, 12:30 — 2:00 PM, December 2, 2020 (#39), 2020
Oil on canvas
6h x 11w in

This painting practice is both a representation of this disconnect and a coping mechanism for dealing with it. It has helped me carry on through the cognitive exhaustion from all the video calls and the anxiety of self-isolation while the sirens howled through the streets around me. I know that isolation is a privilege, but for me it is a necessary one: aggressive treatment for an aggressive cancer has left me with a number of impairments and mostly invisible disability. As an immunocompromised person, I generally have to be careful, as an average flu (despite the flu shot) leaves me in bed for a month. Facing COVID-19, I have had to be exceptionally vigilant to self-isolate.

Like many people who have lived through cancer treatment,

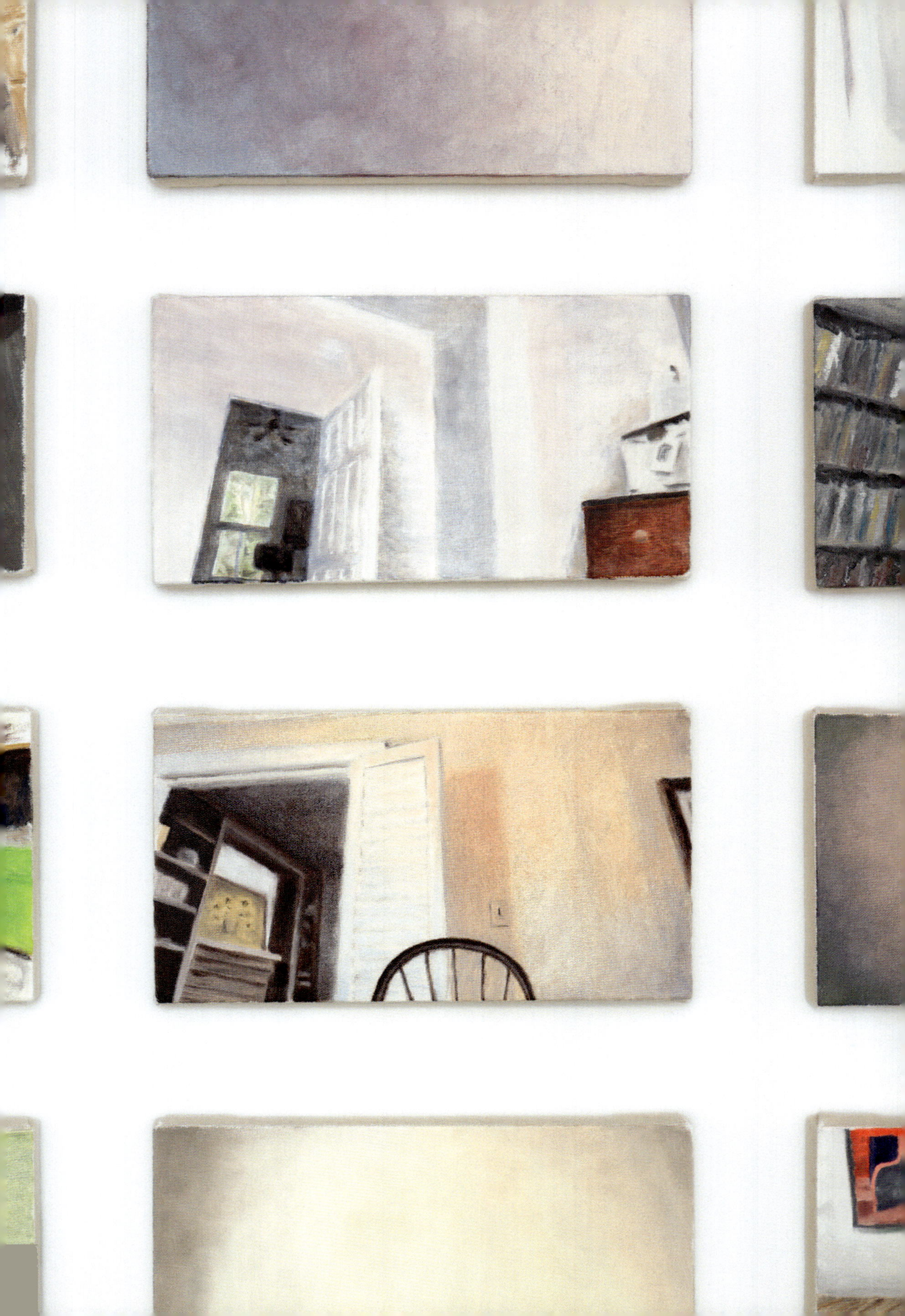

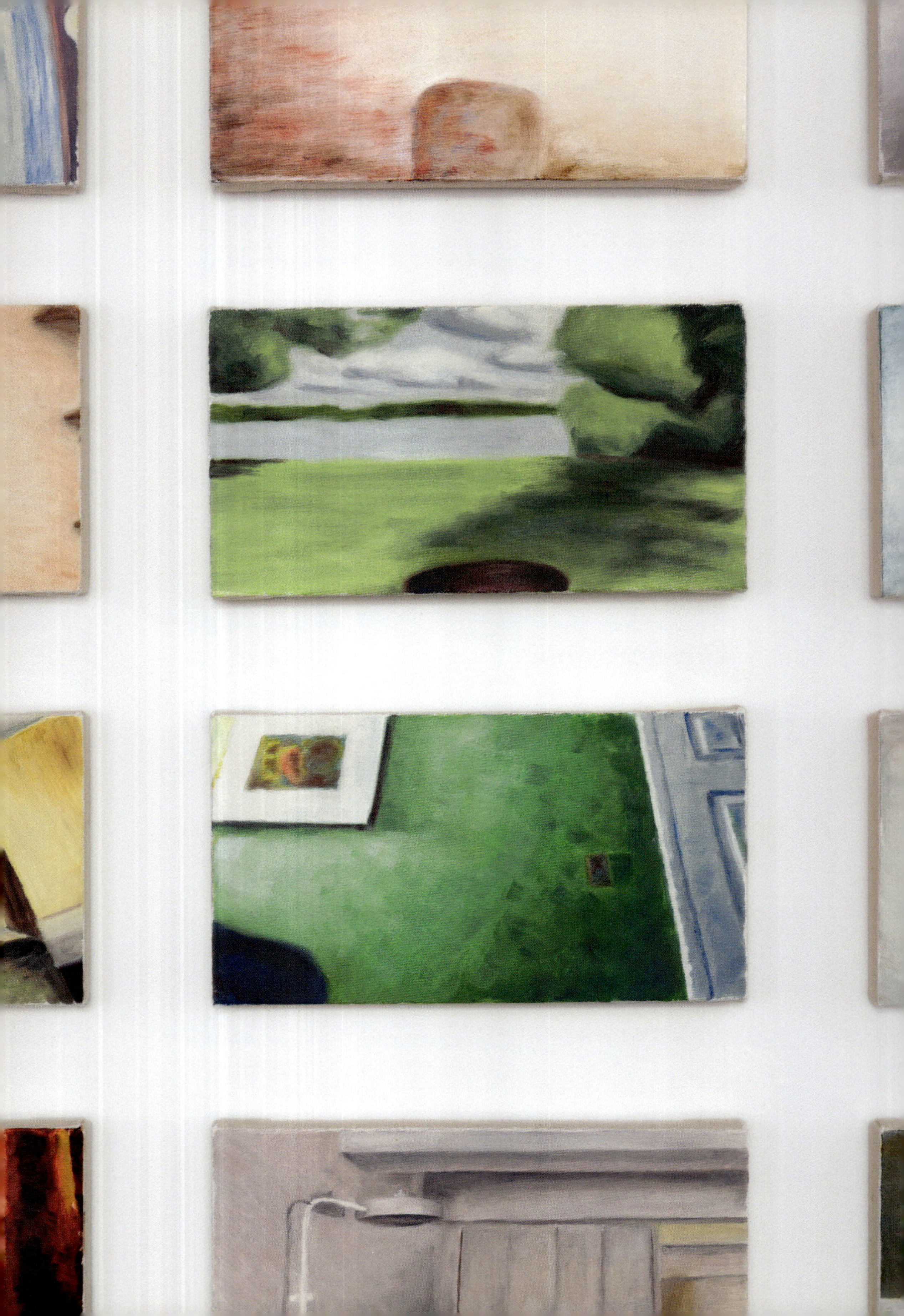

I experience memory loss. These paintings track digital experiences that are likely to slip away and be forgotten. If possible, I try to complete each painting in the time span of the video call from which it was sourced. Still, when the demands of my call keep me from painting, I let myself complete the paintings after the fact. Like the bad lenses, compression algorithms, and choppy connections that mediate these interactions, my memory often fills in the gaps. Sometimes the off-white colors that are transformed to light pinks, yellows, and cool blues are a result of digital glitches, but sometimes they are the result of my glitchy memory.

Painting these canvases has been an attempt to log memory, as much as it is also a practice of self-care and preservation. Time has shifted in the absence of regular routine, and I have lost track of what day it is. These paintings help me keep track of digital experiences that are more likely to slip away and be forgotten. They are memory, and memorials. But I am not memorializing a particular event; rather I am memorizing how unmemorable each is.

I don't announce that I'm painting, though you can tell if you watch me, and at this point, many people know about the project and expect that I will be. About half of the images are of rooms of people I know pretty well, and the other half are random people on large group calls; I chose not to paint any of the rooms of my students. Sometimes I worry that people will be aggrieved that I have used an image of their space, but everyone who has seen an image of their room has been really pleased. I wonder why it seems to resonate so much. Is it because memorializing an aspect of this strange time is comforting? Maybe it is because they have spent so much time in that space but

Facing page:

Detail view of Zoom Paintings installation

Center top:
FaceTime from Stephen Mandiberg, 8:52 — 10:03 PM, November 30, 2020 (#38), 2020
Oil on canvas
6h x 11w in

Center bottom:
Michael + Ying, 11:00 — 1:00 PM, February 27, 2021 (#50), 2021
Oil on canvas
6h x 11w in

have never seen it as it is? Or because when painted, it becomes somewhere else — some people have said that the paintings seem to be only partly recognizable as their spaces.

Many of the images are from large group calls, where people do not know I am painting; these include artist talks, panels, and many varieties of faculty meetings. The Zoom room is a strange liminal space between private and public. We are conducting our public business as if it were a workplace, a classroom, public plaza. And yet each of us is in our private spaces. There is an essay about Zoom, Habermas, and the public sphere for someone else to write; I'm busy painting.

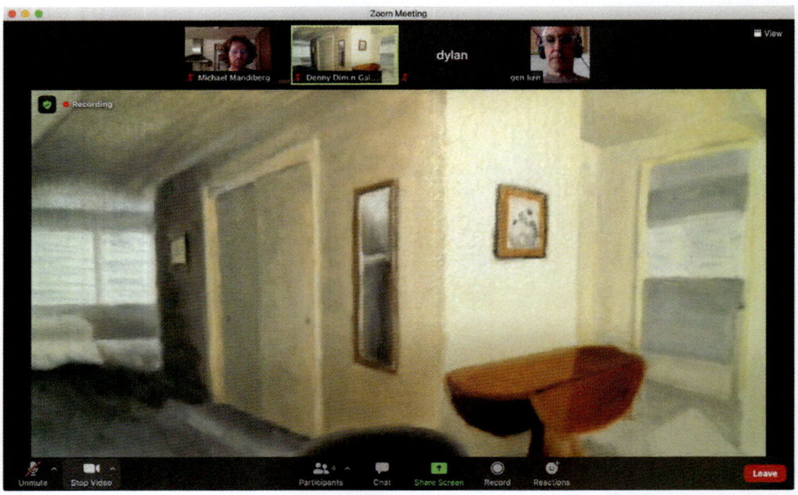

Exhibition documentation from Michael Mandiberg, ***The Zoom Paintings***, Denny Dimin Gallery (virtual), November 12 - 25, 2020

Eventually, when we can safely meet in person again, I intend to show these paintings in a grid that resembles Zoom's gallery view. In the months (or years) until then, I will be sharing these paintings through Zoom itself. In November 2020 Denny Dimin Gallery presented ten of these paintings—in their Zoom meeting room. To see them, viewers could visit this meeting room during gallery hours; the webcam showed each painting for a day before it was exchanged with the next one. Viewers were welcome to stay in the room as long as they liked and return as often as they wanted. The empty room seems to offer people a respite. It certainly functions that way for me when I make them.

I am composing this text in January 2021, nearly a year after the pandemic began. I'm sitting at my improvised work space in London, about to get on a Zoom call. I'm here with my partner who is at the start of a PhD at University College London. We have been in quarantine or lockdown for most of the time since we arrived in October. Before COVID-19, I was planning to visit them in London but not to be here the whole time; this summer we decided that neither of us wants to do this alone, regardless of which side of the ocean we were on. It took some time to come to that decision, but I think it was the right one. It's not like I'm thriving here, but things would be way worse if I were alone in Brooklyn. What a strange time to travel to another country, where I know almost no one, am prohibited by public health policy from meeting anyone, and even if the lockdown eases, no one is particularly interested in making new friends.

Like New York City in April, London is currently the global epicenter of the pandemic. Like then, the city is quiet, save the wailing ambulance sirens. The US vaccination campaign started slowly and ran out of supplies last week; the UK campaign seems steady, though the government here hasn't provided much information about the time line. The newest South African variant is both more lethal and more resistant to the vaccines; no one has been able to confirm it yet, but it is clear that the UK variant is too. I'm starting to get cancellation announcements for summer events. I look forward to the day when we will be able to gather in person, and I will be able to stop making these paintings, though the end seems further in the future with each passing week.

Zoom Paintings

Plates

Office Hours: Introduction to Deletion Policy on Wikipedia [...] CSI Chapter Meeting, 2:00 — 4:30 PM, April 16, 2020 (#1), 2020

Oil on canvas
9h x 12w in

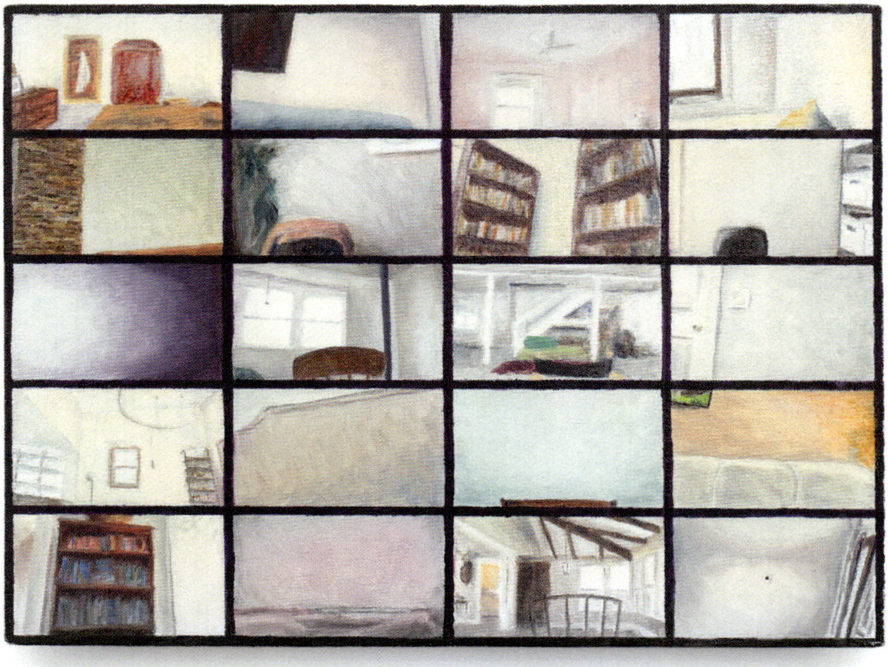

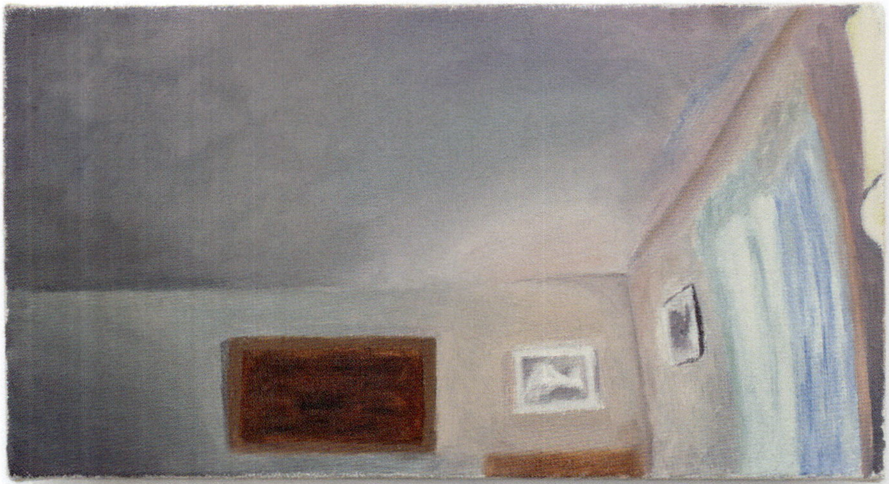

PSC-CUNY Action I, 3:00 —
4:00 PM, June 23, 2020 (#13),
2020

Oil on canvas
6h x 11w in

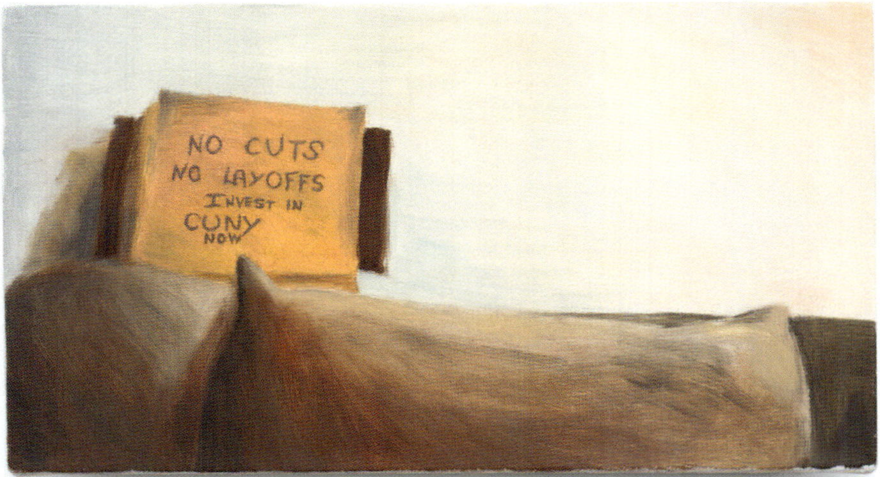

PSC-CUNY Action II, 3:00 — 4:00 PM, June 23, 2020 (#14), 2020

Oil on canvas
6h x 11w in

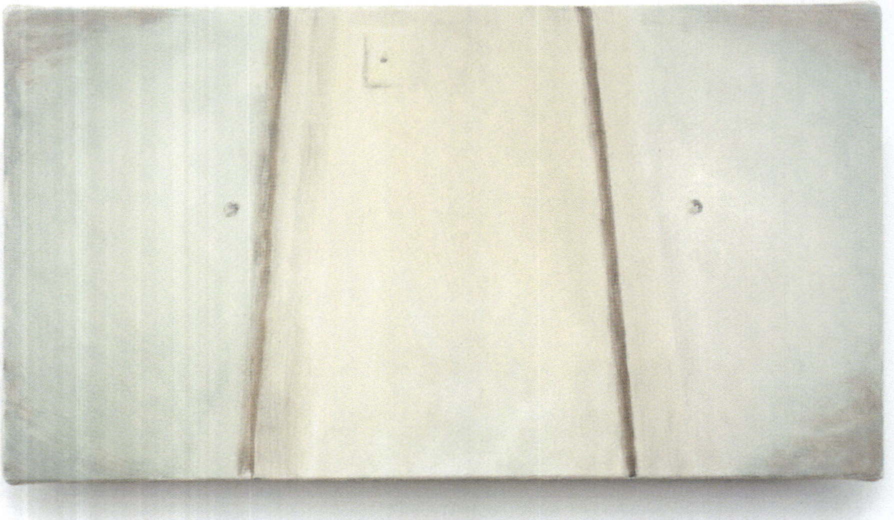

JCOPE Comprehensive Ethics Seminar Training, 10:00 AM — 12:00 PM, August 05, 2020 (#20), 2020

Oil on canvas
6h x 11w in

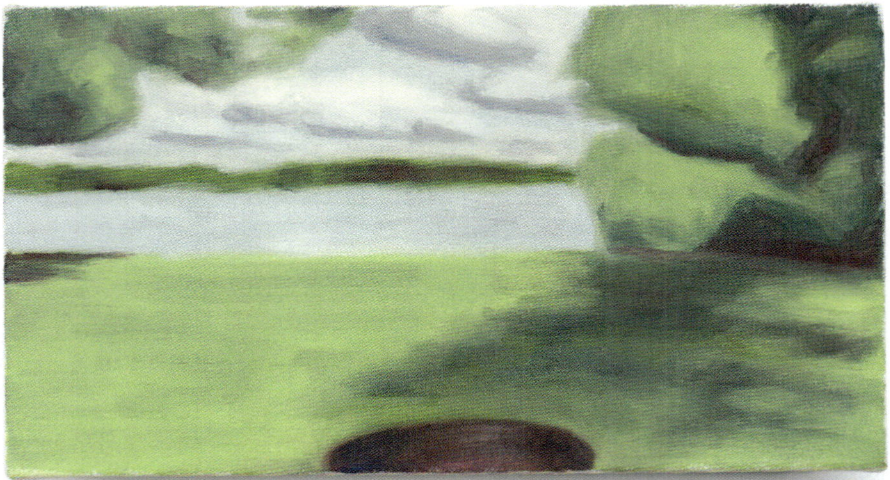

Sara wiknic, 3:00 — 4:00 PM, August 16, 2020 (#23), 2020

Oil on canvas
6h x 11w in

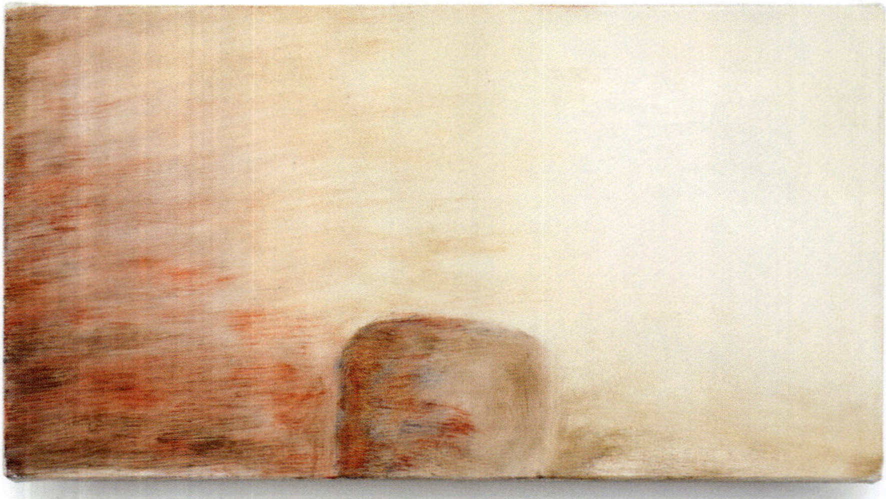

PSC Demands Justice, 6:30 — 8:00 PM, August 26, 2020 (#26), 2020

Oil on canvas
6h x 11w in

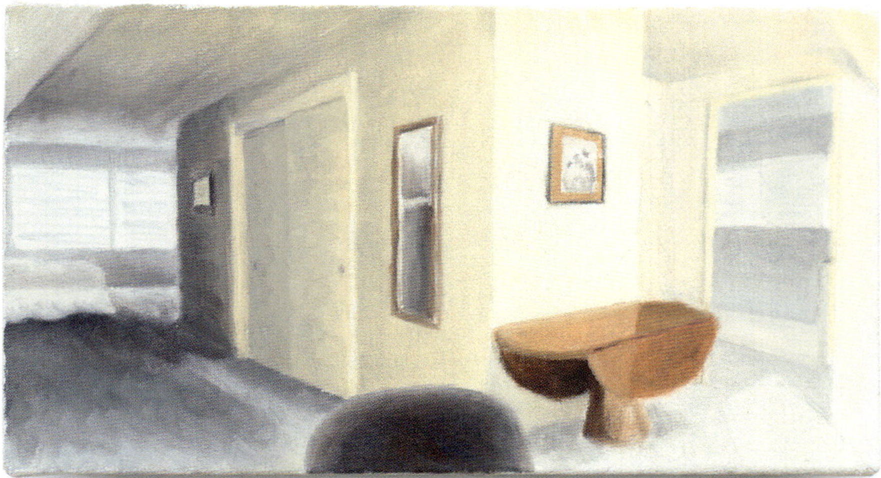

National Phonebank with Artists for Biden/Harris, 6:00 — 8:00 PM, September 20, 2020 (#27),
2020

Oil on canvas
6h x 11w in

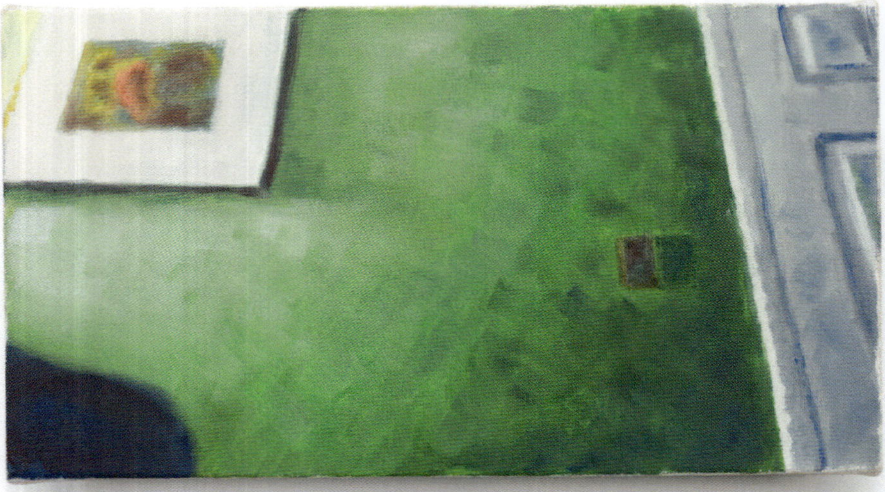

Town Hall, "Concerns About the College's Direction," 6:30 — 8:30 PM, October 29, 2020 (#30),
2020

Oil on canvas
6h x 11w in

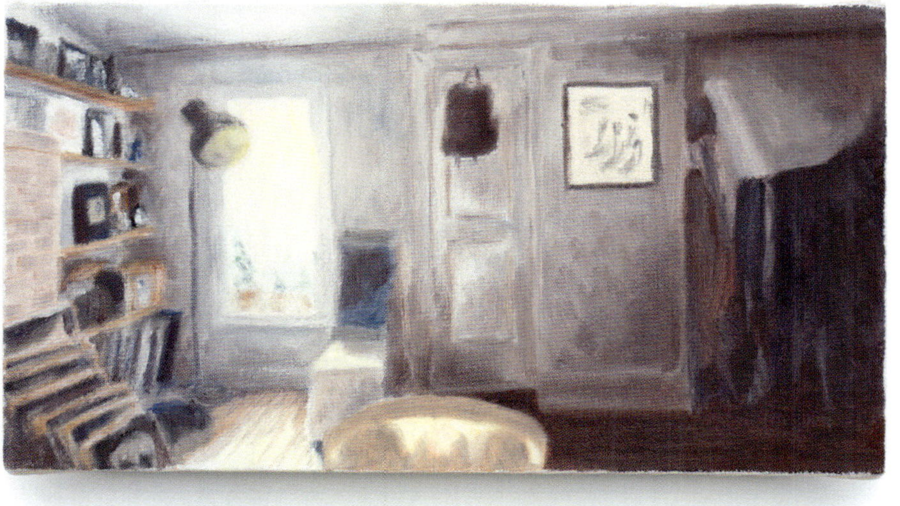

Michael + Taylor re: Artnet, 5:30 — 6:30 PM, November 4, 2020 (#32), 2020

Oil on canvas
6h x 11w in

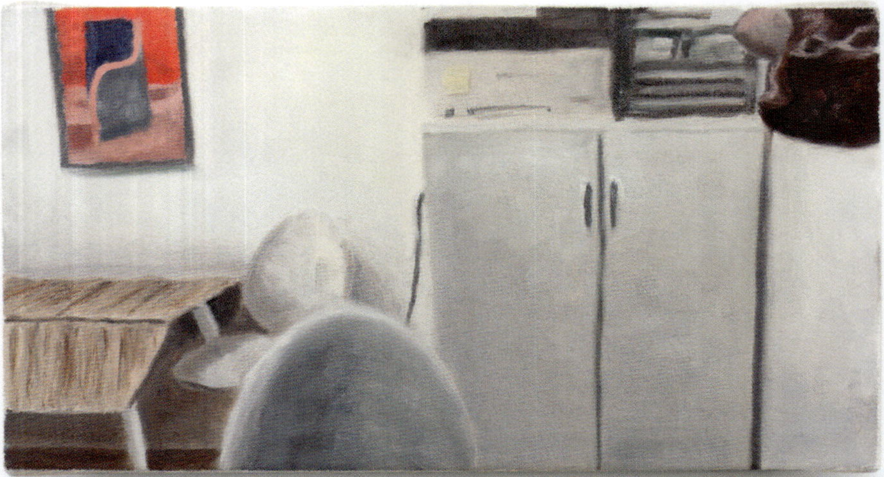

Press Preview for DDG show, 6:00 — 8:00 PM, November 10, 2020 (#34), 2020

Oil on canvas
6h x 11w in

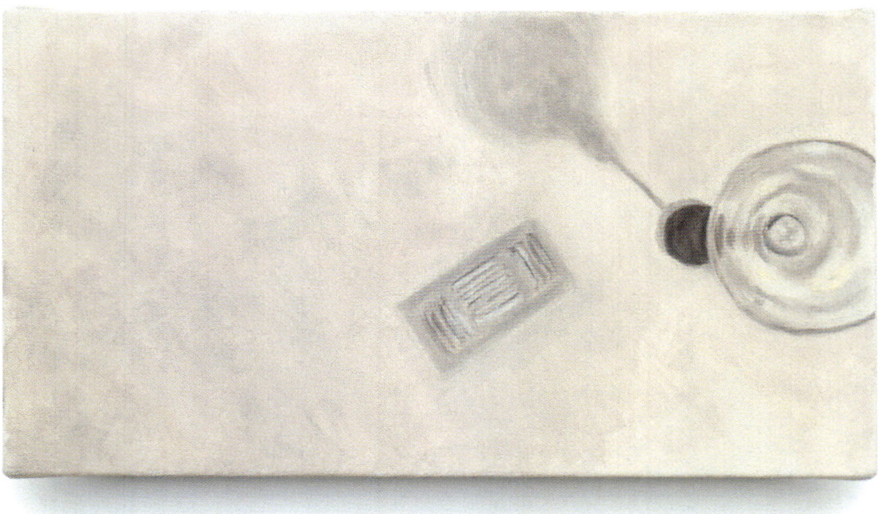

FaceTime from Stephen Mandiberg, 8:52 — 10:03 PM, November 30, 2020 (#38), 2020

Oil on canvas
6h x 11w in

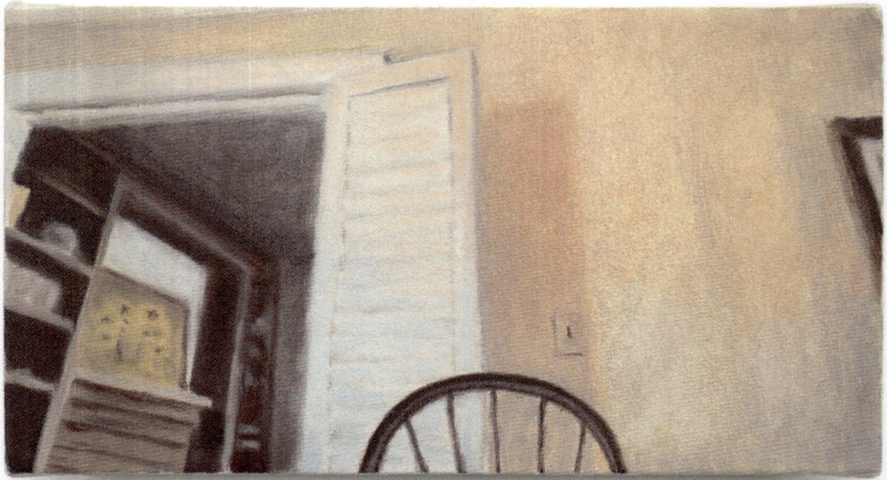

ITP advisory meeting, 12:30 — 2:00 PM, December 2, 2020 (#39), 2020

Oil on canvas
6h x 11w in

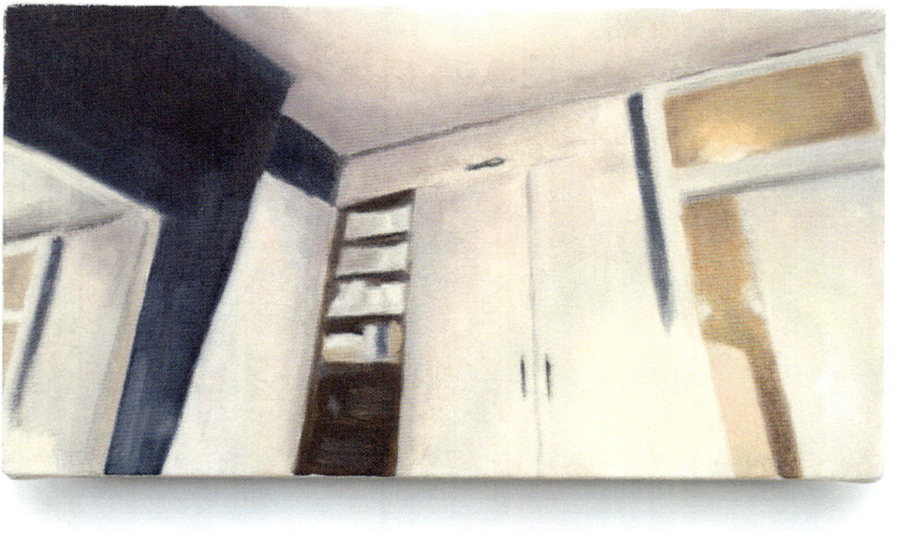

FaceTime from Alec Hall, 8:15 — 9:05 PM, December 5, 2020 (#40), 2020

Oil on canvas
6h x 11w in

What is White Privilege?, 3:00 — 5:00 PM, December 7, 2020 (#41), 2020

Oil on canvas
6h x 11w in

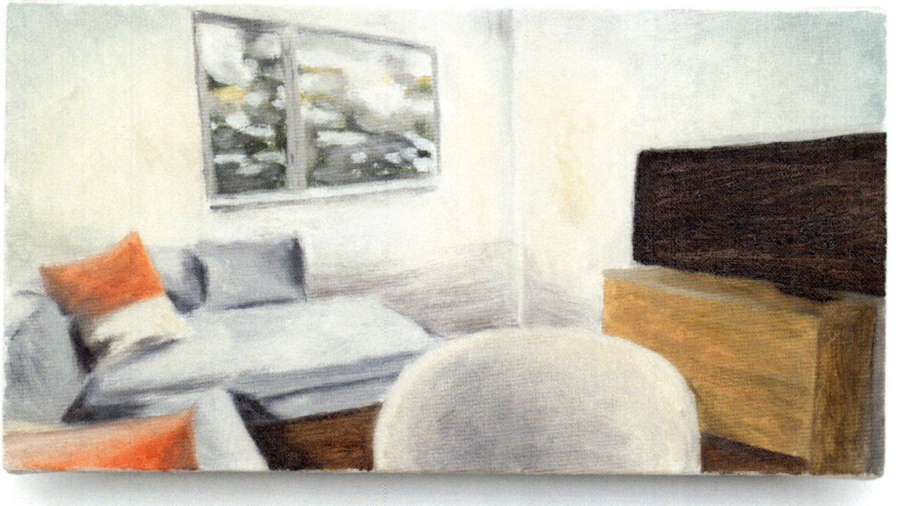

2020 NYFA Fellows and Finalist Celebration 12/09, 5:00 — 7:00 PM, December 9, 2020 (#42),
2020

Oil on canvas
6h x 11w in

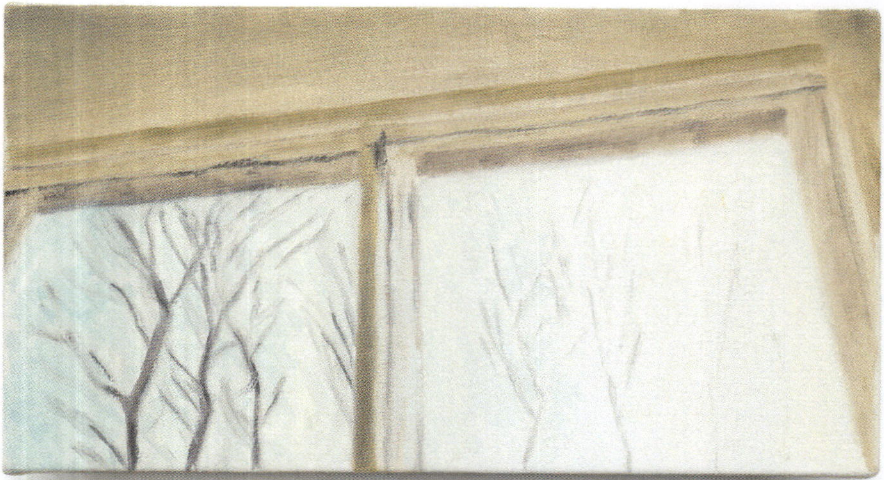

Art + Feminism, Phase 2 part 1a, 6:00 — 9:00 PM, December 15, 2020 (#43), 2020

Oil on canvas
6h x 11w in

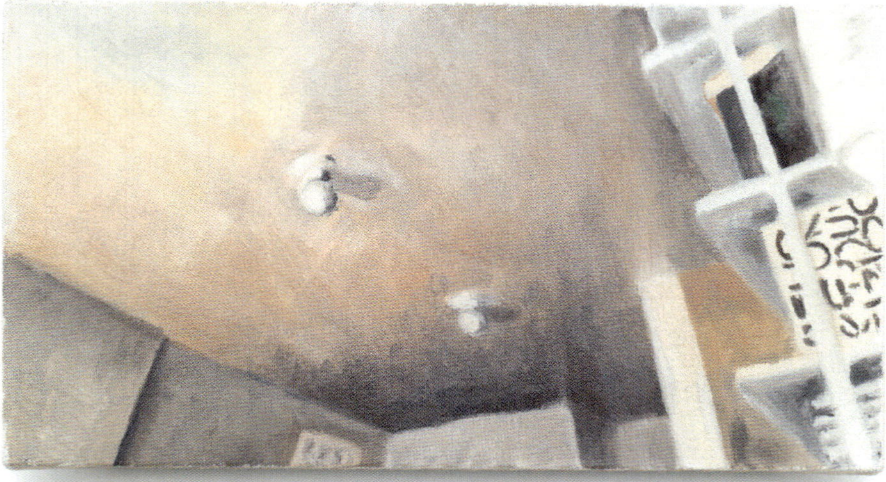

A+F Board Planning Conversation (Comm, Committee, etc), 2:30 — 4:30 PM, January 5, 2021 (#44), 2021

Oil on canvas
6h x 11w in

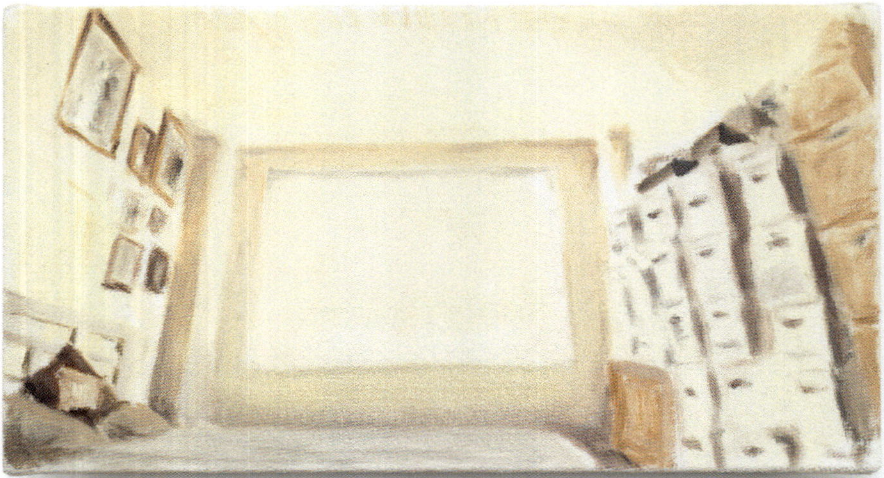

Wikipedia training for Hrag's group, 6:30 — 7:30 PM, January 31, 2021 (#46), 2021

Oil on canvas
6h x 11w in

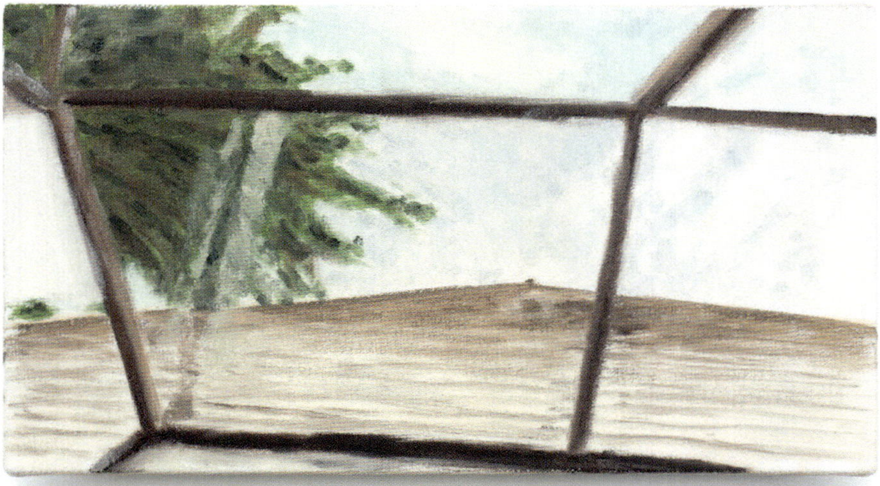

FaceTime from Joseph Mandiberg, 8:52 — 10:03 PM, March 8, 2020 (#53), 2021

Oil on canvas
6h x 11w in

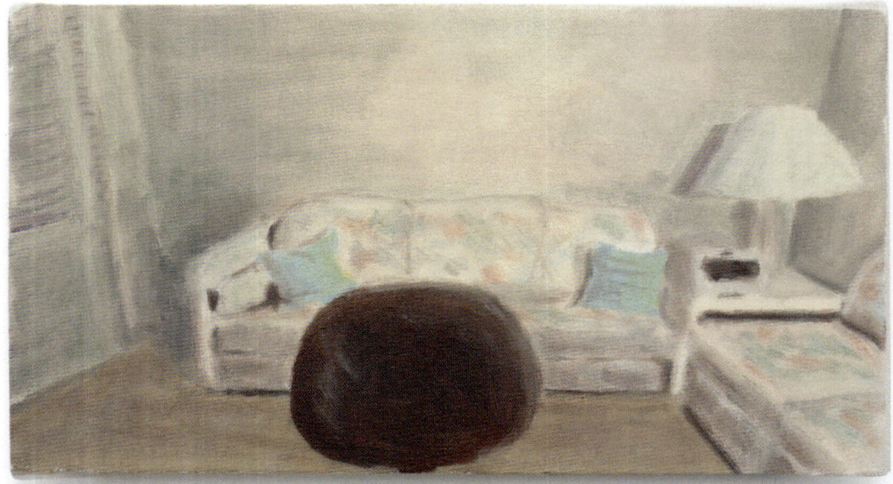

Presidential Referendum on CSI Governance Plan, 7:30 — 9:15 PM, March 11, 2021 (#54), 2021

Oil on canvas
6h x 11w in

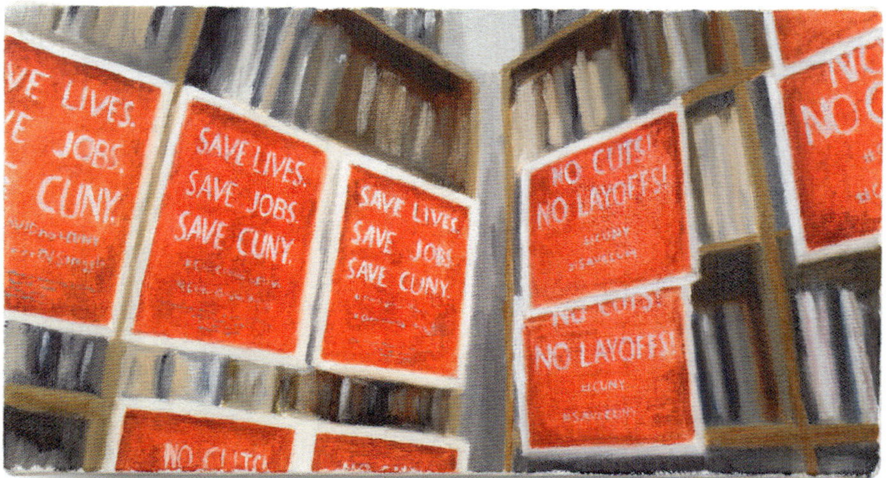

Emergency CUNY Budget Rally Online TOMORROW!, 9:00 — 10:00 PM, April 1, 2021 (#59), 2021

Oil on canvas
6h x 11w in

CSI "Re-entry" Town Hall, 7:00 — 8:30 PM, April 20, 2021 (#61),
2021

Oil on canvas
6h x 11w in

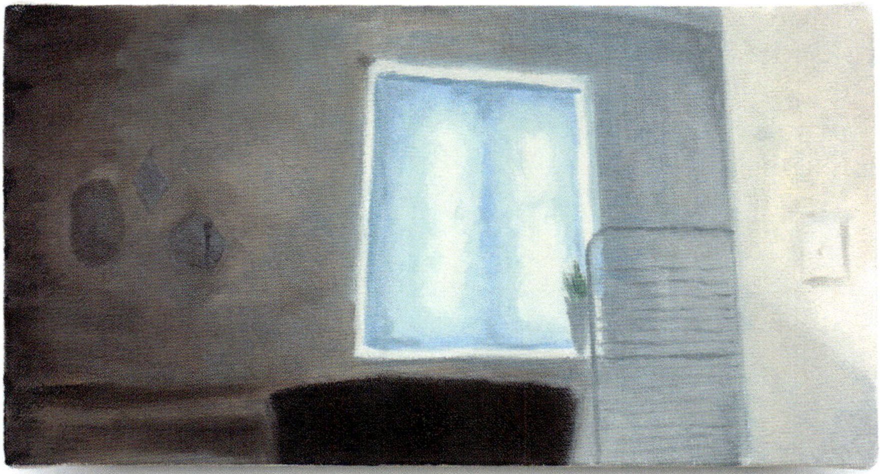

Public Scholarship workshop, 5:00 — 6:30 PM, April 23, 2021 (#62), 2021

Oil on canvas
6h x 11w in

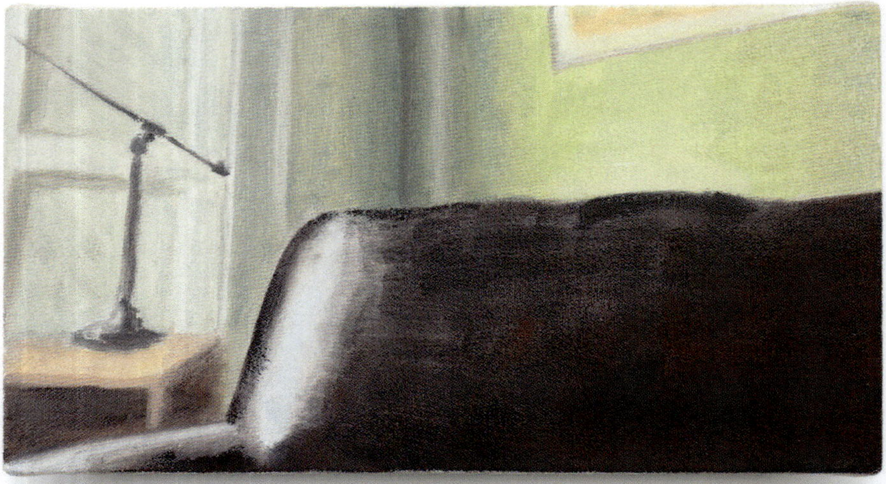

ITP Advisory Committee meeting,
3:00 — 4:30 PM, May 11, 2021
(#64), 2021

Oil on canvas
6h x 11w in

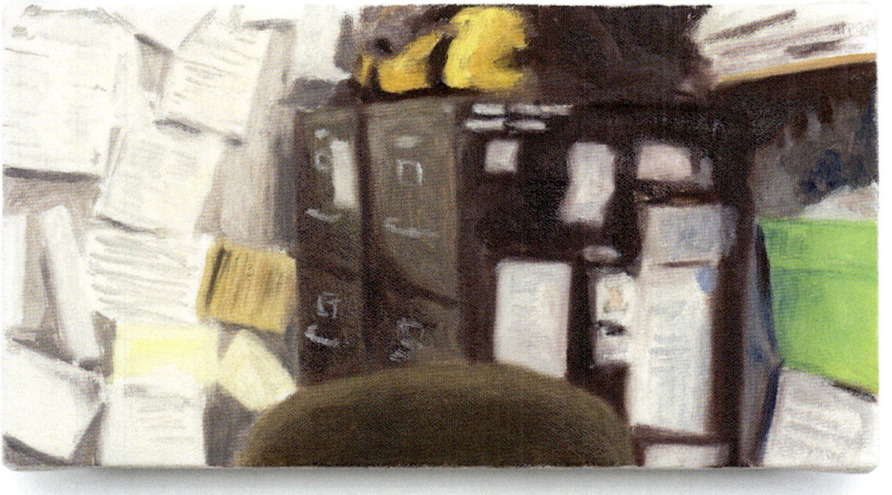

Park Slope Food Coop General Meeting, 7:00 — 9:00 PM, June 30, 2021 (#65), 2021

Oil on canvas
6h x 11w in

Michael Mandiberg

Lives in Brooklyn, New York

Born 1977 in Detroit, Michigan

Education

2003	MFA in Photography and Media, California Institute of the Arts
2000	BA in English (with honors) & Visual Arts, *magna cum laude*, Brown University
1998–2000	Photography Studies including Degree Project, Rhode Island School of Design
1998	Fine Arts Studies, San Francisco Art Institute, Semester Exchange

Selected Projects

2020 - present	*Zoom Paintings*
2019 - present	*Live Study*
2018	*Postmodern Times*
2017	*Quantified Self Portrait*
2008–16	*FDIC Insured*
2015	*Print Wikipedia*
2014 - present	Art+Feminism (co-founder)
2012 - present	New York Arts Practicum
2009	*Burned Books*
2007	*Real Costs*
2006	*Oil Standard*
2005	*IN Network*, with Julia Steinmetz
2005	*All Haiku, All the Time*
2002	*The Essential Guide to Performing Michael Mandiberg*
2001	*AfterSherrieLevine.com*
2001	*Shop Mandiberg*

Selected Solo Exhibitions

2021	*Timeframe,* Denny Dimin Gallery, New York
2020	*The Zoom Paintings*, Denny Dimin Gallery, New York, online exhibition
2019	*Live Study,* Whitney Museum of American Art, online artport exhibition
2017	*New Work,* Denny Gallery, New York

Workflow, Los Angeles County Museum of Art (LACMA)

2016 *FDIC Insured*, Art-in-Buildings Financial District Project Space, New York

Print Wikipedia: From Aachen to Zylinderdruckpresse, Import Projects, Berlin

In the Stacks: Print Wikipedia, Arizona State University Art Museum and Library, Tempe

2015 *From Aaaaa! to ZZZap!*, Denny Gallery, New York

2010 *The Great Recession, Walter Feldman Gallery*, Pacific Northwest College of Art, Portland

2004 *First Person*, Centro de la Imagen, Mexico City

First Person, Sala de Arte Publico Siquieros, Mexico City

Selected Group Exhibitions

2021 *The Society of Individuals*, Museum of Contemporary Art, Busan, Korea

2020 *HeK Net Works*, HeK, Basel, October 13 - November 10, 2020, online exhibition

Biennale für aktuelle Fotografie, Mannheim, Germany

Journey into a Living Being, Kunstraum Kreuzberg, Berlin

Algotaylorism, La Kunsthalle Mulhouse, Mulhouse, France

2019 *Hyperemployment*, Aksioma at the Museum of Graphic Arts in Ljubljana

Museum of Capitalism, Kellen Gallery at Parsons School of Design, New York

24/7, Somerset House, London (curated by Sarah Cook)

WEB-RETRO, SeMA Buk-Seoul Museum of Art

2018 *This Site is Under Revolution*, Moscow International Biennale for Young Art, MMOMA

The Current // Truth, The Current Museum, New York, NY

2017 *POST FAIL*, Fotomuseum Winterthur, Winterthur

Analog Currency, The Mistake Room, Los Angeles

Opening Exhibition, Museum of Capitalism, Oakland

2016 *update_6 Biennial* / New Technological Art Award (NTAA), Ghent

Pandora's Box, Musée d'Art Moderne de la Ville de Paris

I Wanted to Be a Photographer, Fundació Foto Colectania, Barcelona

Sight & Sound, Eastern Bloc, Montreal

2014 *Share This!: Appropriation After Cynicism*, Denny Gallery, New York

2013 *DECENTER: An Exhibition on the Centenary of the 1913 Armory Show*, Abrons Arts Center, New York

Curitiba International Biennial, Curitiba. Curated by Teixeira Coelho and Ticio Escobar

2012 *Art, Environment, Action!*, Sheila C. Johnson Design Center, New York

2011 *The Oil Show*, Hartware MedienKunstVerein (HMKV), Dortmund

Identités précaires, Jeu de Paume, Paris

2009 *The End of Oil*, Exit Art, New York

Histoires à l'ère numérique, Plug In, Basel

the future is not what it used to be, Postmasters, New York

2008 *Ours: Democracy in the Age of Branding*, Sheila C. Johnson Design Center, New York

Feedback: Beyond Light Bulbs, Eyebeam, New York

2006 *When Artists Say We*, Artists Space, New York

2005 *Rhizome ArtBase 101*, New Museum of Contemporary Art, New York

Just Do It, Lentos Kunstmuseum, Linz

2003 *Tirana Biennale 03*, National Gallery of the Arts, Tirana

onscreen_4: Echo Sparks, Ars Electronica, Linz

2002 *Go Public!*, transmediale, Berlin

2001 *Multiple Personalities*, Haines Gallery, San Francisco

International Media Art Award, Zentrum für Kunst und Medientechnologie (ZKM), Karlsruhe

Selected Reviews

2020 Sophie Haigney, "'The Zoom Paintings' are dispatches from a strange year," *The Economist*, December 3

Taylor Dafoe, "'It's Memorializing How Unmemorable It Is': Artist Michael Mandiberg on Painting Melancholy Portraits on Zoom," *Artnet*, November 10

2019 Casey Lesser, "Watch an Artist Paint Every Assistant They've Ever Had," *Artsy*, July 30

2018 Ligaya Mishan, "In Literature, Who Decides When Homage Becomes Theft?" *The New York Times Magazine*, October 8

2017 Régine Debatty, "Failed Banks, Quantified Self," *We Make Money Not Art*, September 7

Alex Greenberger, "MoMA Announces Fourth Annual Art+Feminism Wikipedia Edit-a-Thon," *ARTNews*, February 8

2016 Seph Rodney, "Where Do Banks Go When They Die?" *Hyperallergic*, September 29

Mark Singer, "Making Art with Failed Banks," *The New Yorker*, September 22

Tina Rivers Ryan, "Market Research," *Art In America*, August 26

Jessica Binsch "Einmal die Wikipedia ausdrucken, bitte!," *Süddeutsche Zeitung*, May 28

Julia Halperin, "Don't bank on it: New York artist's memorial to financial failures comes to Wall Street," *The Art Newspaper*, May 10

Talia Lavin, "A Feminist Edit-a-Thon Seeks to Reshape Wikipedia," *The New Yorker*, Mar 11

2015 Jennifer W. Leung, "Michael Mandiberg," *Artforum*, October 2015

	Jennifer Schuessler, "Moving Wikipedia from Computer to Many, Many Bookshelves," *The New York Times,* June 17
	Anna Russell, "'Edit-a-Thon' Volunteers Revamp Female Artists' Wikipedia Profiles," *The Wall Street Journal,* March 9
	Jennifer Schuessler, "MoMA to Host Wikipedia Editing Marathon," *New York Times,* March 6
2014	Robin Cembalest, "101 Women Who Got Wikipedia Pages This Week," *ARTNews*, February 6
2011	Josephine Bosma, *Nettitudes: Let's Talk Net Art*, Institute of Network Cultures/NAi Publishers
2009	Edward A. Shanken, *Art and Electronic Media*, Phaidon Themes and Movements Series
2006	Reena Jana and Mark Tribe, *NewMedia Art*, Taschen/Basic Art, pp. 13-14
	Joline Blais and Jon Ippolito *At the Edge of Art*, Thames and Hudson, pp. 116-117
2003	Rachel Greene, *World of Art Series: Internet Art*, Thames and Hudson, pp. 184-185, 191
2001	Achy Obejas, "Identity for Sale," *Chicago Tribune*, February 6
	Matthew Mirapaul, "A Market for Flotsam and Jetsam as Performance Art," *The New York Times*, February 5

Selected Grants and Awards

2021	Elizabeth Foundation for the Arts, Studio Membership
2020	NYFA Fellowship, Digital/Electronic Arts
	Silver Art Projects, World Trade Center Residency
2018	Residency, Fonderie Darling, Montreal
2017	Shortlist, Kunstbulletin and HeK Basel net-based award
	Residency, 18th Street Art Center, Los Angeles
2016	LACMA Art + Technology Award
	MacDowell Colony Fellowship
2014-20	8 Wikimedia Foundation Grants for Art+Feminism, and Print Wikipedia
2014	100 Leading Global Thinkers, Foreign Policy Magazine
2011	Mellon Fellowship, Center for the Humanities, CUNY Graduate Center
2008, 2013	2 Residencies, The Banff Centre
2006-10	Resident, Research Fellow, Senior Fellow, Eyebeam
2006	Rhizome.org Commission
2002-06	3 Turbulence Project Commissions with funds from the Jerome Foundation, and NYC DCA

Selected Collections

21c Museum, Louisville, KY

Arizona State University, Tempe, AZ

Los Angeles County Museum of Art (LACMA)

The Espace Multimedia Gantner, Bourgogne, France

The Stedelijk Museum, Amsterdam, The Netherlands

Whitney Museum of American Art, New York, NY, USA

Teaching

2011–present	The Graduate Center, CUNY, Doctoral Faculty
2004–present	College of Staten Island (CUNY), Professor (current rank)

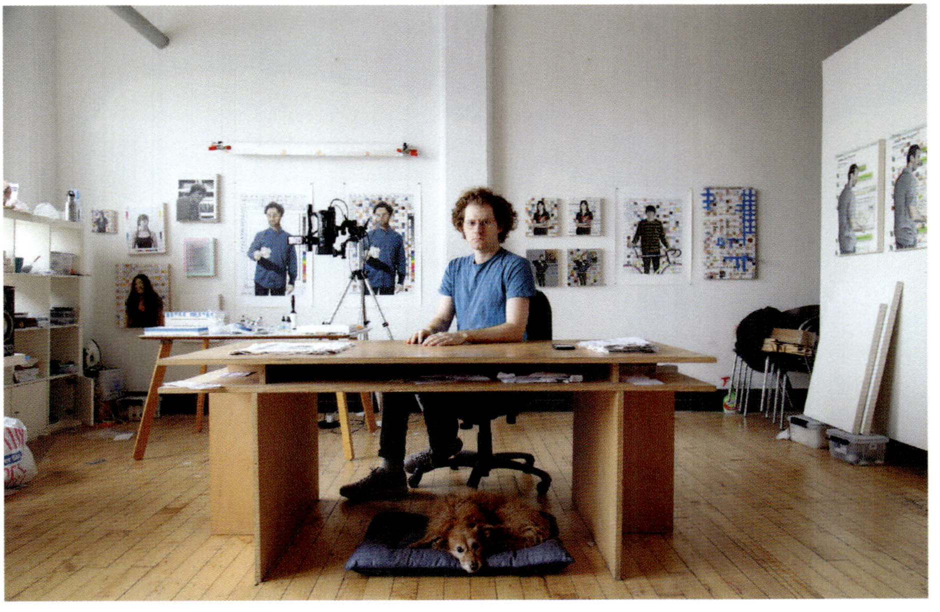

Michael Mandiberg and Freddie Merckx, 2019.

Published on the occasion of the exhibition
Michael Mandiberg: *Timeframe* at Denny Dimin Gallery

First edition published in 2021
Edition of 300

Denny Dimin Gallery
39 Lispenard Street, New York, NY 10013

Michael Mandiberg's essay was originally published in *Social Text*, Issue 149, 2021

Acknowledgments
Jacqueline Mabey
Joseph and Linda Mandiberg
Christiane Paul
danilo machado
College of Staten Island, City University of New York
Fonderie Darling
Bakehouse Art Complex
Elizabeth Foundation for the Arts
Peter Rostovsky
William Powhida
Sean Fader

Design by
Kuldeep Singh
Brooklyn, NY

ISBN 978-0-578-30632-2

DENNY DIMIN
GALLERY